SYMBIOTIC PLANET

By the Same Author

What Is Sex? Lynn Margulis and Dorion Sagan

Slanted Truths: Essays on Gaia, Symbiosis and Evolution
Lynn Margulis and Dorion Sagan

***Five Kingdoms: An Illustrated Guide to the Phyla of
Life on Earth, Third Edition***
Lynn Margulis and Karlene V. Schwartz

What Is Life? Lynn Margulis and Dorion Sagan

***Garden of Microbial Delights: A Practical Guide to the
Subvisible World*** Dorion Sagan and Lynn Margulis

***Symbiosis in Cell Evolution: Microbial Communities in the
Archaen and Proterozoic Communities, Second Edition***

***The Illustrated Five Kingdoms: A Guide to the Diversity of
Life on Earth*** Lynn Margulis, Karlene V. Schwartz and
Michael Dolan

Diversity of Life: The Five Kingdoms

***Concepts of Symbiogenesis: A Historical and Critical
Study of the Research of Russian Botanists***
Edited by Mark McMenamin and Lynn Margulis

***Environmental Evolution: Effects of the Origin and
Evolution of Life on Planet Earth***
Edited by Lynn Margulis and Lorraine Olendzenski

Origins of Sex: Three Billion Years of Genetic Evolution
Lynn Margulis and Dorion Sagan

Symbiosis as a Source of Evolutionary Innovation
Edited by Lynn Margulis and Rene Fester

Mystery Dance: On the Evolution of Human Sexuality
Lynn Margulis and Dorion Sagan

***Microcosmos: Four Billion Years of Evolution from Our
Microbial Ancestors*** Lynn Margulis and Dorion Sagan

SYMBIOTIC PLANET

A New Look at Evolution

LYNN MARGULIS

BASIC
BOOKS

A Member of
The Perseus Books Group

The Science Masters Series is a global publishing venture consisting of original science books written by leading scientists and published by a worldwide team of twenty-six publishers assembled by John Brockman. The series was conceived by Anthony Cheetham of Orion Publishers and John Brockman of Brockman Inc., a New York literary agency, and developed in coordination with Basic Books.

••••••••••••••

The Science Masters name and marks are owned by and licensed to the publisher by Brockman Inc.

••••••••••••••

Published by Basic Books,
A Member of the Perseus Books Group.

••••••••••••••

FIRST EDITION

••••••••••••••

A CIP catalog record for this book is available from the Library of Congress.
ISBN 0-465-07271-2

98 99 00 01 ❖/RRD 10 9 8 7 6 5 4 3 2 1

CONTENTS

···

List of Illustrations   vi

    Prologue   1

1   Symbiosis Everywhere   5

2   Against Orthodoxy   13

3   Individuality by Incorporation   33

4   The Name of the Vine   51

5   Life from Scum   69

6   Animal Sex   87

7   Ashore   105

8   Gaia   113

Appendix   129

Notes   131

Index   137

About the Author   147

LIST OF ILLUSTRATIONS

1   Prokaryotic and eukaryotic cells compared   7

2   SET (serial endosymbiosis theory) phylogeny   31

3   Spirochetes become undulipodia   35

4   Five-kingdom hand   53

5   *Mixotricha paradoxa,* a protoctist   63

6   Ediacaran biota of the Late Proterozoic era   96

7   Mating protists   100

Figure credits: 1. Christie Lyons; 2, 6 Laszlo Meszoly; 3, 5 Kathryn Delisle; 4, 7 Dorion Sagan

....

> Time does go on -
> I tell it gay to those who suffer now -
> They shall survive -
> There is a sun -
> They don't believe it now – (1121)*

Mom, what does the Gaia idea have to do with your symbiotic theory?" asked my son Zach, age seventeen, after work one day. No longer an aspiring politician, now a disillusioned aide to a legislator at the State House in Boston, he had just returned home from an exhausting attempt to draft old-people's-home legislation for one of his two absentee bosses.

"Nothing," I immediately responded, "or at least nothing as far as I'm aware." I have been pondering his question ever since. The book you hold in your hands attempts to provide the answer. The two major scientific ideas that I have worked on all my professional life, serial endosymbiosis theory (SET) and Gaia, and the relation of one to the other, form its central theme.

Zach's question, how symbiosis jibes with Gaia, was neatly answered by a wisecrack of a wonderful former stu-

---

*All chapter epigraphs are quotations from Emily Dickinson (1830–1886); numbered in T.H. Johnson, editor, *The Complete Book of Poems of Emily Dickinson,* (Little Brown and Company, 1955).

dent of mine named Greg Hinkle, now a professor at the University of Massachusetts, at South Dartmouth. Before receiving his Ph.D., Greg knew and taught that symbiosis is simply the living together in physical contact of organisms of different species. Partners in symbiosis, fellow symbionts abide, literally touching each other or even inside each other, in the same place at the same time. The concept "Gaia," an old Greek name for Mother Earth, postulates the idea that the Earth is alive. The Gaia hypothesis, proposed by the English chemist James E. Lovelock, is that aspects of the atmospheric gases and surface rocks and water are regulated by the growth, death, metabolism, and other activities of living organisms. Greg quips, "Gaia is just symbiosis as seen from space": all organisms are touching because all are bathed in the same air and the same flowing water. The reasons I think Greg is correct are detailed in the pages that follow.

If this book teaches you about symbiosis and Gaia theory in the context of radically new views of life, it is only because of four lucky facts: first, Zach's question; second, the contribution of Dorion Sagan to the quality of my thinking and writing[1]; third, Lois Byrnes queried, reorganized, and restructured this manuscript with visionary honesty and meticulous artistic taste[2]; last, the appropriate insistence of William Frucht, of Basic Books, on more focused organization and less self-indulgent narrative. The pleasure of working with such an intellectually curious and properly critical editor continues.

This book is about planetary life, planetary evolution, and the ways our views of them are changing. If there is a subtext, it concerns exploration, specifically scientific exploration, and the many quirks and agenda that can nurture or block it. Many circumstances conspire to extinguish scientific discoveries, especially those that cause discomfort about our culture's sacred norms. As a species, we cling to the familiar, comforting conformities of the mainstream. However, "convention" penetrates more deeply than we tend to admit. Even if we lack a proper name for and knowledge of the history of any specific philosophy or thought style, all of us are embedded in our own safe "reality." Our outlooks shape what we see and how we know. Any idea we conceive as fact or truth is integrated into an entire style of thought, of which we are usually unaware. Call the cultural constraints "trained incapacities," "thought collectives," "social constructions of reality." Call the dominating inhibitions that determine our point of view whatever you wish. They affect all of us, including scientists. All are saddled with heavy linguistic, national, regional, and generational impediments to perception. Like those of everyone else, the scientist's hidden assumptions affect his or her behavior, unwittingly restricting thought.

One widely held unstated assumption is the great chain of being. Defining the venerable position of humans as the exact center of the universe in the middle of the chain of being below God and above rock. This anthropocentric idea dominates religious thought, even that of those who claim to reject religion and to replace it with a scientific worldview. For the Greeks, the chain joined a panoply of gods at the top to, in descending order, men, women, slaves, animals, and vegetables. A substratum of rocks and minerals occupied the lowest link. The Judeo-Christian version allowed slight modification: people, above animals, were positioned a little lower than the angels. Man, of course, was indisputably and obviously superseded by the Almighty.[3]

These ideas are rejected as obsolete nonsense by the scientific worldview. All beings alive today are equally evolved. All have survived over three thousand million years of evolution from common bacterial ancestors. There are no "higher" beings, no "lower animals," no angels, and no gods. The devil, like Santa Claus, is a useful myth. Even the "higher" primates, the monkeys and apes, in spite of their name (*primate* comes from Latin, *primus*, "first") are not higher. We *Homo sapiens sapiens* and our primate relations are not so much special entities: rather, we are newcomers on the evolutionary stage. Human similarities to other life-forms are far more striking than the differences. Our deep connections, over vast geological periods, should inspire awe, not repulsion.

As a species, we still fear the eccentric in our views of ourselves. Despite or perhaps because of Darwin, as a culture we still don't really understand the science of evolution. When science and culture conflict, culture always wins. Evolutionary science deserves to be much better understood. Yes, humans have indeed evolved, but not just from apes or even from other mammals. We evolved from a long line of progenitors, ultimately from the first bacteria.

Most evolution occurred in those beings we dismiss as "microbes." All life, we now know, evolved from the smallest life-forms of all, bacteria. We need not welcome this fact. Microbes, especially bacteria, are touted as enemies and denigrated as germs. Microbes, in fact, are any live beings—algae, bacteria, yeast, and so forth—seen more accurately with a microscope than as smudges or scum with the naked eye. My claim is that, like all other apes, humans are not the work of God but of thousands of millions of years of interaction among highly responsive microbes. This view is unsettling to some. To some it is frightening news from science, a rejectable source of information. I find it fascinating: it spurs me to learn more.

....

CHAPTER I

SYMBIOSIS EVERYWHERE

A Bee his burnished Carriage
Drove boldly to a Rose —
Combinedly alighting —
Himself — (1339)

Symbiosis, the system in which members of different species live in physical contact, strikes us as an arcane concept and a specialized biological term. This is because of our lack of awareness of its prevalence. Not only are our guts and eyelashes festooned with bacterial and animal symbionts, but if you look at your backyard or community park, symbionts are not obvious but they are omnipresent. Clover and vetch, common weeds, have little balls on their roots. These are the nitrogen-fixing bacteria that are essential for healthy growth in nitrogen-poor soil. Then take the trees, the maple, oak, and hickory. As many as three hundred different fungal symbionts, the mycorrhizae we notice as mushrooms, are entwined in their roots. Or look at a dog, who usually fails to notice the symbiotic worms in his gut. We are symbionts on a symbiotic planet, and if we care to, we can find symbiosis everywhere. Physical contact is a nonnegotiable requisite for many differing kinds of life.

Practically everything I work on now was anticipated by unknown scholars or naturalists. One of my most important scientific predecessors thoroughly understood and explained the role of symbiosis in evolution. The University of Colorado anatomist Ivan E. Wallin (1883–1969) wrote a fine book arguing that new species originate through symbiosis. *Symbiogenesis,* an evolutionary term, refers to the origin of new tissues, organs, organisms—even species—by establishment of long-term permanent symbiosis. Wallin never used the word *symbiogenesis,* but he entirely understood the idea. He especially emphasized animal symbiosis with bacteria, a process he called "the establishment of microsymbiotic complexes" or "symbionticism." This is important. Although Darwin entitled his magnum opus *On the Origin of Species,* the appearance of new species is scarcely even discussed in his book.[1]

Symbiosis, and here I fully agree with Wallin, is crucial to an understanding of evolutionary novelty and the origin of species. Indeed, I believe the idea of species itself requires symbiosis. Bacteria do not have species.[2] No species existed before bacteria merged to form larger cells including ancestors to both plants and animals. In this book I will explain how long-standing symbiosis led first to the evolution of complex cells with nuclei and from there to other organisms such as fungi, plants, and animals.

That animal and plant cells originated through symbiosis is no longer controversial. Molecular biology, including gene sequencing, has vindicated this aspect of my theory of cell symbiosis. The permanent incorporation of bacteria inside plant and animal cells as plastids and mitochondria is the part of my serial endosymbiosis theory that now appears even in high school textbooks. But the full impact of the symbiotic view of evolution has yet to be felt. And the idea that new species arise from symbiotic mergers among members of old ones is still not even discussed in polite scientific society.

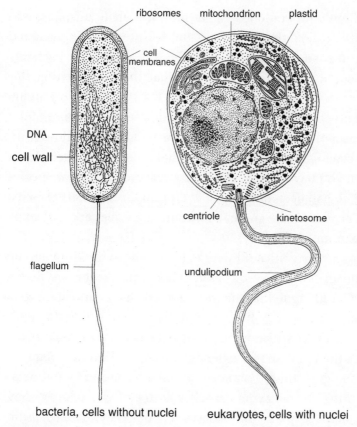

ribosomes   mitochondrion   plastid

cell
membranes

DNA

cell wall

centriole   kinetosome

flagellum   undulipodium

bacteria, cells without nuclei   eukaryotes, cells with nuclei

FIGURE I

Prokaryotic and eukaryotic cells compared

Here is an example. I once asked the eloquent and personable paleontologist Niles Eldredge whether he knew of any case in which the formation of a new species had been documented. I told him I'd be satisfied if his example were drawn from the laboratory, from the field, or from observations from the fossil record. He could muster only one good example: Theodosius Dobzhansky's experiments with *Drosophila*, the fruit fly. In this fascinating experiment, populations of fruit flies, bred at progressively hotter temperatures, became genetically separated. After two years or so the hot-bred ones could no longer produce fertile offspring with their cold-breeding brethren. "But," Eldredge quickly added, "that turned out to have something to do with a parasite!" Indeed, it was later discovered that the hot-breeding flies lacked an intracellular symbiotic bacterium found in the cold breeders. Eldredge dismissed this case as an observation of speciation because it entailed a microbial symbiosis! He had been taught, as we all have, that microbes are germs, and when you have germs, you have a disease, not a new species. And he had been taught that evolution through natural selection occurs by the gradual accumulation, over eons, of single gene mutations.

Ironically, Niles Eldredge is author with Stephen Jay Gould of the theory of "punctuated equilibrium." Eldredge and Gould argue that the fossil record shows evolution to be static most of the time and to proceed suddenly: rapid change in fossil populations occurs over brief time spans; stasis then prevails for extended periods. From the long view of geological time, symbioses are like flashes of evolutionary lightning. To me symbiosis as a source of evolutionary novelty helps explain the observation of "punctuated equilibrium," of discontinuities in the fossil record.

Among the only other organisms besides fruit flies in which species have been seen to originate in the laboratory are members of the genus *Amoeba* and symbiosis was involved. Symbiosis is a kind, but not the notorious kind, of Lamarckianism. "Lamarckianism," named for Jean Baptiste Lamarck, who the French claim was the first evolutionist, is often dismissed as "inheritance of acquired characteristics." In simple Lamarckianism, organisms inherit traits induced in their parents by environmental conditions, whereas through symbiogenesis, organisms acquire not traits but entire other organisms, and of course, their entire sets of genes! I could say, as my French colleagues often have, that symbiogenesis is a form of neo-Larmarckianism. Symbiogenesis is evolutionary change by the inheritance of acquired gene sets.[3]

Living beings defy neat definition. They fight, they feed, they dance, they mate, they die. At the base of the creativity of all large familiar forms of life, symbiosis generates novelty. It brings together different life-forms, always for a reason. Often, hunger unites the predator with the prey or the mouth with the photosynthetic bacterium or algal victim. Symbiogenesis brings together unlike individuals to make large, more complex entities. Symbiogenetic life-forms are even more unlike than their unlikely "parents." "Individuals" permanently merge and regulate their reproduction. They generate new populations that become multiunit symbiotic new individuals. These become "new individuals" at larger, more inclusive levels of integration. Symbiosis is not a marginal or rare phenomenon. It is natural and common. We abide in a symbiotic world.

In Brittany, on the northwest coast of France, and along beaches bordering the English Channel is found a strange sort of "seaweed" that is not seaweed at all. From a distance it is a bright green patch on the sand. The patches slosh around, shimmering in shallow puddles. When you pick up the green water and let it slip through your fingers you notice gooey ribbons much like seaweed. A small hand lens or low-power microscope reveals that what looked like seaweed are really green worms. These masses of sunbathing green worms, unlike any seaweed, burrow into the sand and effectively disappear. They were first described in the 1920s by an Englishman, J. Keeble, who spent his summers at Roscoff. Keeble called them "plant-animals" and diagrammed them splendidly in the color frontispiece of his book, *Plant-Animals*. The flatworms of the species *Convoluta roscoffensis* are all green because their tissues are packed with *Platymonas* cells; as the worms are translucent, the green color of *Platymonas*, photosynthesizing algae, shows through. Although lovely, the green algae are not merely decorative: they live and grow, die and reproduce, inside the bodies of the worms. Indeed they produce the food that the worms "eat". The mouths of the worms become superfluous and do not function after the worm larvae hatch. Sunlight reaches the algae inside their mobile greenhouses and allows them to grow and feed themselves as they leak photosynthestic products and feed their hosts from the inside. The symbiotic algae even do the worm a waste management favor: they recycle the worm's uric acid waste into nutrients for themselves. Algae and worm make a miniature ecosystem swimming in the sun. Indeed, these two beings are so intimate that it is difficult, without very high-power microscopy, to say where the animal ends and the algae begin.

Such partnerships abound. Bodies of *Plachobranchus*, snails, harbor green symbionts growing in such even rows they appear to have been planted. Giant clams act as living gardens, in which their bodies hold algae toward the light. *Mastigias* is a man-of-war type of medusoid that swims in the Pacific Ocean. Like myriad small green umbrellas, *Mastigias*[4] medusoids float through the light beams near the water's surface by the thousands.

Similarly, freshwater tentacled hydras may be white or green, depending on whether or not their bodies are packed with green photosynthetic partners. Are hydras animals or plants? When a green hydra is permanently inhabited by its food producing partners (called *Chlorella*), it is hard to tell. Hydras, if green, are symbionts. They are capable of photosynthesizing, of swimming, of moving, and of staying put. They have remained in the game of life because they become individuals by incorporation.

We animals, all thirty million species of us, emanate from the microcosm. The microbial world, the source and wellspring of soil and air, informs our own survival. A major theme of the microbial drama is the emergence of individuality from the community interactions of once-independent actors.

I love to gaze on the daily life struggles of our nonhuman planetmates. For many years Lorraine Olendzenski, my former student, now at the University of Connecticut, and I have videographed life in the microcosm. More recently we have worked with Lois Byrnes, the vivacious former associate director of the New England Science Center in Worcester, Massachusetts. Together we and a fine group of U MASS students make films and videos that introduce people to our microbial acquaintances.

Ophrydium, a pond water scum that, upon close inspection, seems to be countable green "jelly ball" bodies is an example of emergent individuality that we recently discovered in Massachusetts and redescribed. Our films show these water balls with exquisite clarity. The larger "individual" green jelly ball is composed of smaller cone-shaped actively contractile "individuals." These in turn are composite: green *Chlorella* dwell inside ciliates, all packed into rows. Inside each upside-down cone are hundreds of spherical symbionts, cells of *Chlorella*. *Chlorella* is a common green alga; those of *Ophrydium* are trapped into service for the jelly ball community. Each "individual organism" in this "species" is really a group, a membrane-bounded packet of microbes that looks like and acts as a single individual.

A nutritious drink called kefir consumed in the Caucasus Mountains is also a symbiotic complex. Kefir contains grainy curds the Georgians call "Mohammed pellets." The curd is an integrated packet of more than twenty-five different kinds of yeast and bacteria. Millions of individuals make up each curd. From such interactive bodies of fused organisms new beings sometimes emerge. The tendency of "independent" life is to bind together and reemerge in a new wholeness at a higher, larger level of organization. I suspect that the near future of *Homo sapiens* as a species requires our reorientation toward the fusions and mergers of the planetmates that have preceded us in the microcosm. One of my ambitions is to coax some great director into producing evolutionary history as the microcosmic image in IMAX or OMNIMAX, showing spectacular living relationships as they form and dissolve.

Now and throughout Earth's history, symbioses, both stable and ephemeral, have prevailed. Stories about this kind of evolution deserve broadcasting.

··

CHAPTER 2

AGAINST ORTHODOXY

The hills erect their Purple Heads
The Rivers lean to see
Yet Man has not of all the Throng
A Curiosity (1688)

I cannot remember any emotional pain more severe than that of my thirteenth year. No professional disappointment or romantic rejection has ever plunged me more into misery and inactivity. In secret exercise of my perceived rights as a person of free will I snuck out of the University of Chicago eighth-grade laboratory school, with its vastly inferior pool of potential boyfriends, and returned to the huge public high school where I had decided I belonged. I refused to stay another day in that lab school, where everything was so familiar and algebra was so hard.

I was living in my parents' lovely South Shore Drive apartment and decided that running away was the only solution. Of course, I had no money, nowhere to go, and a rigid schedule of classes and duties. When the unfeasibility of running away became obvious and the days lengthened and the weather grew sharply cold, I concocted a plan. Having entered the U of C laboratory school in fourth grade in the September class, the only one they had, I knew of course that

I had been put back a semester. My friends in public school were one half a year ahead of me. As my misery peaked in November or December, as the first semester of eighth grade was coming to a close, my plan firmed up. I would abandon the intricacies of algebra forever and enroll, with my old public school friends, in the ninth grade in Hyde Park High School, five thousand students strong. After a hideous session in which my father made it abundantly clear that I would do no such thing, I realized that my plan had to go underground. One fine low-sun day in early February with a glorious feeling of release from the grips of responsibility, I played hooky. I took a bus and found the huge anonymous office in the confused and policed urban high school at Sixty-third Street. I signed up for ninth grade, the level of school for which I considered myself highly qualified, and told the school officials, when they asked, that I had attended elementary school at the University of Chicago but, after missing the autumn semester, had recently arrived with my parents from out of town.

For some twelve weeks I simply went to all my assigned classes. I most enjoyed Mrs. Kniazza, a superb teacher of Spanish. I performed as a model student. My parents, of course, had no reason to think that I was not in the lab school U of C laboratory high school on a daily basis and I had no reason to disabuse them. Some time late in the spring I received a notice. Hyde Park High School had been informed when they sent for a copy of my elementary school academic record that I had not completed eighth grade at U of C. They concluded I had no authority whatsoever to be attending Hyde Park. I was called to the Hyde Park High School principal's office for interrogation. No, I had never finished U of C High, but why should I have bothered? I asserted. From kindergarten at O'Keefe Elementary School I had been a February entrant and I was returning to my O'Keefe graduate friends, who were now at Hyde Park High. Joining my old classmates, I was simply returning to the status quo ante. Fury hit the fan when the high school administration realized that my parents had no idea that I was not in the lab school; when I had told them that I was leaving I hadn't admitted that my parents didn't know. Of course my parents had not noticed the missing tuition bill.

Many teary sessions followed in and out of school. I don't remember whether my father or I came up with the solution. In the end we worked it out when we asked ourselves how adolescents from foreign schools with incommensurate evaluation systems are properly placed in the U.S. secondary level education. We requested I be given the tests in math, English, history, and humanities for foreign high school students. I easily passed at the ninth grade level. I won the battle. I was permitted to complete ninth grade at Hyde Park, where I enjoyed a far wider choice of boyfriends.

But I lost the war. After two years of public high school my academic advisers told me, when, as an "early entrant" college student, I reentered the University of Chicago, that I had declined in mathematical ability, that my vocabulary had diminished, and that, in general, I was a poorer student at the end of tenth grade than I had been halfway through eighth. When, in the spring of 1954, I finally left the urban racial misery of Hyde Park to attend The College (as the U of C was called, even though they accepted students at a very early age) I was primed, after a two-year lapse, to become a fine student again. Back where I belonged, according to my anxious parents, I was poised to meet the very best of handsome, smart, and eligible young men. The Sagan years followed.

At age fourteen I was lucky indeed to be accepted into the University of Chicago's special early entrant program. Although three and a half years later I graduated with many acquisitions, including a liberal arts degree and a husband, by far the most lasting was a thoroughgoing, finely nurtured critical skepticism. I cherish my University of Chicago education for its central teaching: one must always strive to distinguish bullshit from authenticity.

My fellow student, the budding astronomer Carl Sagan, was nearly five years my senior. Tall, handsome, with a shock of brown-black hair, and exceedingly articulate, even then he was full of ideas. I literally ran into him one day as I was bounding up the steps of Eckhart Hall, the mathematics building. Nineteen-year-old Sagan at that time was poised to launch his astronomical career. He was a graduate student of physics, I but a fast-moving, enthusiastic, ignorant girl.

I was a scientific ignoramus. Carl, and especially his gift of gab, fascinated me. Already he seemed to be a polished professional. From our first meeting he shared with me, and with anyone else who would listen, his keen understanding of the vastness of time and space. As president of the astronomy club, a published journalist, and public speaker, he showed everyone around him how we ignorant chickens could energetically join in the scientific enterprise. His love for science was contagious.

Nonetheless, our passionate foray into loveland had a rocky start and equally abrasive end. My father hated his arrogance and my mother was always suspicious of his self-centered character. We traveled south to Mexico and east to New Jersey. We went together many places and broke up several times. I even made a tape recording for myself enumerating why marrying him would be a stupid, self-destructive move. Yet on June 6, 1957, when I was nineteen, we enjoyed a marvelous wedding. My mother, master of the elegant, had organized only the best. His mother, Rachel Sagan, although of course she was present throughout the festivities, that day sent me a telegram. It read: "From a Bachelor to a Mrs. in only one week." Graduation ceremonies for my magnificent Robert M. Hutchins legacy, the no-major, no-elective University of Chicago liberal arts degree, had been held a week before our wedding. Hutchins (1899–1977), innovator of liberating higher education, ascended to the presidency of the University of Chicago where he remained for twenty-two years at the age of thirty. Although he had emmigrated west by the time I arrived, to his genius and reformed curriculum I owe my education.

Convinced, in part by Carl, that inheritance phenomena would ultimately succumb to a unique chemical explanation, I thought the science of genetics would give us the best clue to how evolution works. At nineteen I unknowingly followed George Gaylord Simpson's lead.[1] Simpson had written that evolution isn't the most important subject because he studied it; rather, he studied evolution because it is the most important science.

That September, just days before the launch of the Soviet satellite *Sputnik,* I accompanied Carl to Yerkes Observatory, Williams Bay, Wisconsin. Yerkes, the University of Chicago's astronomy department, was situated some ninety miles away from the great city on the lake. There Carl worked as a graduate student, one of only four in planetary science, then a nearly unheard of discipline. Already he had begun his search for planets habitable by life.

I too applied to a graduate program. Shocked that they accepted my philosophy liberal arts no-major as a real bachelor's degree, I became a master's degree candidate ninety miles northwest, at the University of Wisconsin at Madison. It was 1958. Pregnant, and sleepy in class, I studied cell biology and genetics in two departments. I began as a teaching fellow where I was most needed, in the department of zoology in the liberal arts college. My other work was in the genetics department in the agriculture school. I was taught general genetics and then population genetics by my master teacher, James F. Crow. I adored Crow's general genetics course; it changed my life. I knew when I left the University of Chicago I wanted to study genetics, but after Crow's class I knew I wanted to study *only* genetics. I figured the study of genetics is the precise way to reconstruct evolution, the story of early, prehuman, life on Earth. When my eldest son, Dorion, and I were writing our book *"Microcosmos"** I remember going, at his request, into the Boston MBTA Auditorium–Massachusetts Avenue station to see the graffiti scrawled on a worn map in a subway tunnel. The huge black letters queried, "Whence Come Amoebae in Chaos?" I laughed out loud seeing the statement, my focus, in essence my life question, on the dim walls of the dingy underground metro stop.

I still ponder the beginnings of the remote past. What had happened to life itself on Earth in its earliest days? After that first course I felt that the field of population genetics, with its insistence on overly abstract neo-Darwinian concepts such as "mutational load," "fitness," and "coefficients of selection," taught more of a religion than a description of the rules by which real organisms passed on their genes and evolved.

Genes, as everyone knew by then, were centered in the nucleus of each plant and animal cell, where they were passed to offspring, usually unchanged. These genes, we easily learned through detailed problem sets and superb Crow lectures, determined, indeed controlled, the traits of the descendants.

I was no more constitutionally inclined to focus monomaniacally on the cell nucleus than I was to be a satellite wife in a nuclear family; my attentions, like that of many women, were divided. My friend Mary Catherine Bateson describes modern women as "peripheral visionaries." A woman must be almost octopoid in her attentions if she is to survive. Holding the infant in one arm, Bateson points out, she stirs the pot with another, while she watches the toddler. These multiple pressures were not then, nor are they now, wished away by political will and feminist rhetoric.

---

* See page 132.

My work began off-center. I studied the genetic systems others tended to ignore, those residing in cell structures (organelles) outside the nucleus. The data on "cytoplasmic genes" fascinated me from the time I first learned about them. The cytoplasm is the liquid portion of the cell, in which mitochondria, chloroplasts, and other organelles are found.[2] Genes were supposed, then as now, to be centralized in the nucleus. Since cytoplasmic genes were confusing, the experiments establishing their existence were often incompletely described. I was not the first to draw attention to cytoplasmic genes. Indeed, many early workers on cell genetics, or as it was also called at that time, "cytoplasmic heredity," were aware of these genes. Jan Sapp, currently a professor at York University near Toronto, in his *Evolution by Association*, wonderfully describes the intellectual history of this entire subfield of genetics.[3] Cytoplasmic genetics research began in the first decade of the twentieth century, at the same time that nuclear genetics work started. Both lines of inquiry began with the rediscovery of Gregor Mendel's work, which established only the nuclear genes. Mendel, the Bohemian gardening monk, inferred the existence of genes, which he called "factors," by rules of inheritance of alternate traits in garden peas. His work in the 1860s was rediscovered by three different scientists in 1990 long after his death. Mendel is applauded by cliché when he is accused of being the "father of genetics." Genetics researchers, euphoric about their early discoveries of Mendel's nuclear factors (which became nuclear genes), were perplexed by nonnuclear (or cytoplasmic) genetic systems. Boris Ephrussi, a yeast genetics expert and Russian emigré living in France, quipped that "there are two kinds of genetic systems, nuclear and unclear." Of course by "unclear" he meant cytoplasmic.[4]

What began at the margins of the microscopic view of the cell has now moved closer to center stage. The findings of the importance of symbiosis in evolution have forced us to revise the earlier nucleocentric view of evolution as a bloody struggle of animals. That nature may be "red in tooth and claw," casually indifferent to individual suffering, does not preclude the fact that symbiosis, beginning as an uneasy alliance of distinct life-forms, may underlie the origin of major evolutionary novelty. Human and other animal consciousness, as well as other types of biological beauty and complexity, are properties of our coevolving, pointillist bacterial ancestry. Cellular interliving, an infiltration and assimilation far more profound than any aspect of human sexuality, produced everything from spring-green blooms and warm, wet mammalian bodies to the Earth's global nexus. Symbiogenesis three decades later is converting cytoplasmic genetics from a marginal subject to a central one in gene studies.

The abbot Mendel postulated "factors"; only later did others call them nuclear genes. These factors, Mendel theorized, produced the different colors (yellow and green) and textures (wrinkled or smooth) in the sweet peas he grew and crossbred in his monastery kitchen garden. Mendel's purpose, in all likelihood, was to show the ultimate immutability of inheritance and therefore to counter, like some venerable and brilliant version of an oxymoronic modern-day "creation scientist," Charles Darwin's Adam-smashing notion of the mutability of all species. According to a brilliant unpublished manuscript by an amateur historian of science whose name I can't remember from Nassau Island in the Bahamas, Mendel saw no evidence at all that species change and evolve. Red male and white female flowers produced seeds that grew into pink offspring. But the flowers generated by pink parents were just as red or just as white as their grandparents had been. Whatever his motivations, Mendel's factors were correlated with the inheritance of unchanged characters. Furthermore, these hypothetical factors were strictly associated with the behavior of red-staining chromosomes bounded by nuclear membrane. My colleague Jan Sapp has published analyses of Mendel's work similar to those of the unknown woman who sheepishly arrived at my office with sheaves of paper, evidence of years of her work never to see the light of day.

Chromosomes, tiny bodies residing inside the nucleus of animal and plant cells, were known long before the discovery of the structure of deoxyribonucleic acid (DNA) in 1953. By the time I came of scientific age, the chromosomal theory of heredity had already become canonized as truth. The "theory" designation had been discarded and it was taught as fact: genes were "on chromosomes." The evidence was unequivocal that tucked inside the nucleus of tissue cells the hypothetical genes were on the chromosomes. These genes corresponded exactly to Mendel's theoretical factors: they played by the rules and determined whether there were red, white, or pink flowers in plants and comparably inherited characteristics in animals. Evidence for the nuclear location of the trait-determining genes was considered robust enough that the newfound genetic knowledge could be summarized as "the chromosomal basis of heredity."

After the midfifties, mainstream biologists, or "biochemical and biophysical cytologists" as they were then known, found considerable excitement in the search for the detailed "material basis," the actual stuff that make up Mendel's factors. Of what were the red-staining chromosomes composed? What is the chemistry of heredity? At the time science itself, as in some Gothic or science fiction novel, was unveiling the secret of life, the Faustian thrill of such a quest can hardly be overstated. With false starts but ultimately spectacular successes, understanding of the innards of the cell and its nucleus grew. The underlying chemistry of the incessantly active cell was revealed. From food molecules proteins were synthesized and nucleic acid replicated. These chemical activities were the basis of the metabolism of all life.

But an egg was no nucleated bag full of genes. As embryologists and botanists continued to point out, cytoplasmic genes or cytoplasmic factors in egg cells of both plants and animals, yet not inside the nucleus, also exerted control over traits. Factors external to the nucleus were found to be deeply implicated in oxygen breathing and the coloring of leaves.

The genes, in other words, are not necessarily in the nucleus. In both plants and animals some cell genetic factors are dispersed. Since the 1930s, when early biochemical work was done in Germany and England, it has been firmly established in yeast and other fungi that the mitochondria contain their own genes. These little organelles are the sites where oxygen gas from the air reacts with food molecules to yield chemical energy. Green algal and plant cells contain conspicuous green bodies, the chloroplasts, where photosynthesis converts sunlight into usable chemical energy and to food. Chloroplasts, too, have their own genes. This was discovered at the turn of the century by H. De Vries and C. Correns, two plant scientists, who independently rediscovered Mendel's genes. The chloroplast inherited from only one parent, usually the female, apparently determined greenness. The inheritance pattern was nonnuclear.

"From the point of view of heredity, the cytoplasm of a cell can safely be ignored." Even when I first read this statement, confidently uttered by T. H. Morgan, a Columbia professor and key player in the founding of genetics, in 1945, I considered it an arrogant oversimplification.[5] Cell heredity, both nuclear and cytoplasmic, always must be considered for the entire cell, the entire organism.

If Carl played a prominent role in my adolescent conversion to science, more crucial probably was "The College" of the University of Chicago. In my scientific education a critical first step was a year long course of study, Natural Science 2. Instead of textbooks, we students in Nat Sci 2 biological science classes read the writings of the great scientists themselves: Charles Darwin; Gregor Mendel; and the German biologists Hans Spemann, an embryologist active in the first two decades of the twentieth century, and August Weismann, who codiscovered fertilization and postulated the "continuity of the germ plasm." We also read anglophone neo-Darwinists, including the British mathematicians and geneticists G. S. Hardy, J. B. S. Haldane, and R. A. Fisher. Hardy, Haldane, Fisher, and many others developed the mathematic principles of population genetics, a crucial pillar that holds neo-Darwinism aloft. Nat Sci 2 inspired us to consider population genetics, embryology, and many concepts: What is heredity? What links the generations? How do the materials in fused egg and sperm inspire the development of an entire animal? As taught in Nat Sci 2, science was a liberal art, a way of knowing. We were taught how, through science, we could go about answering important philosophical questions. The issues of deep heredity that first consumed me in Nat Sci 2 class inspire me to this day.

The superb science at the University of Chicago, a set of methods that are honest, open, accessible, and energetic, seems hardly to exist in the "technological fix" mentality of today. Science there facilitated the query of profound questions where philosophy and science merge: What are we? Of what are we and the universe made? "Where do we come from? How do we work? I do not doubt that I owe my choice of scientific career to the genius of this "idiosyncratic" education.

Reading the Nat Sci 2 syllabus, I heard the voices of great biologists in my mind's ear: Vance Tartar dissecting the "giant ciliate" *Stentor*; Thomas Hunt Morgan establishing the supremacy of the nucleus; Hermann J. Muller defining life as "mutation, reproduction and the reproduction of mutation," Theodosius Dobzhansky chasing fruit flies in his unending quest to correlate genes, chromosomes, environment, and evolutionary history. Pipe-smoking A. H. Sturtevant, in the Columbia University fly room, sought the chromosomal bases for the factors that determined the characteristics of different kinds of fruit flies. Genetics and evolution, geneticists and evolutionists, even those long dead, exerted a spell over me from my first encounter with their work. The body of coherent science produced by the U.S. school of geneticists in the first half of the twentieth century gave me a sense of the history of biological—especially genetical—thought. From the beginning the need for chemical explanation was clear.

My fascination with evolution began in Nat Sci 2. Theodosius Dobzhansky, still active at Columbia University when I first read his work, wrote that "nothing in biology makes sense except in the light of evolution."[6] Evolution, defined simply as change through time, brings into focus the convoluted history of which we are the living legacy. The study of evolution is vast enough to include the cosmos and its stars as well as life, including human life, and our bodies and our technologies. Evolution is simply *all* of history.

Even as an undergraduate I sensed something was too pat, too reductionistic, too limiting about the idea that genes in the nucleus determine all the characteristics of plants or animals. How could random gene mutations lead to the evolution of flowers and eyes? My skepticism was only confirmed as, following Carl to Wisconsin, I continued my biological work at Madison. Preferring to look directly at living cells rather than grind them up to examine their intrinsic chemistry (metabolism), I became intrigued with chromosomes and other inherited organelles, visible bodies inside cells. I devoted myself to the search for the rules of their transmission.

By 1963, many papers on cytoplasmic factors of eggs showed mysterious genes outside the nucleus. Green female plants crossed with white males in some cases give rise to only green offspring. Yet in those same species, if green male plants are crossed with white females, only white offspring grow from seeds. Why? In the inheritance of nuclear genes the female and male contributions are equal, and it doesn't matter which parent is which. That the egg or the plant cell is not simply a bag that holds the nucleus with its all-important genes was clear to me, as it was to the geneticist predecessors whose work I had read. T. H. Morgan's advice to ignore the cytoplasm seemed to me, even then, simply denial.

Realization that the emphasis on connecting genetics to chemistry had unnecessarily given scientists too narrow a perspective, one overly focused on the nucleus, was my jumping-off point. I studied the work by Ruth Sager and Francis Ryan on cytoplasmic genes and the strange genetic cases of molds collected by the Italian researcher Gino Pontecorvo.[7] Experiments described by these authors showed that two kinds of organelles, membrane-bounded structures inside cells but outside the nucleus, plastids and mitochondria, had significantly affected heredity. The references in these books led me to *The Cell in Development and Heredity*, the 1928 masterpiece by E. B. Wilson.[8] Wilson reviewed early books that talked about the similarity of the cell organelles, the plastids and mitochondria, to free-living microbes. This hint led me to study microbes in the symbiosis literature. As I noted the abundance of symbiotic encounters in nature, especially bacteria living together with, and sometimes inside cells of insects or worms, I became intrigued by the early investigators to whom Wilson referred: I. E. Wallin, K. S. Merezhkovsky, and A. S. Famintsyn. The story is told in Khakhina's superb book, now in English.[9] I suspected they were correct when they hypothesized that nonnuclear cell parts, with their own peculiar heredity, were remnant forms of once free-living live bacteria. It seemed obvious to me that there were double inheritance systems with cells inside cells.

Later I found out that idea had been obvious as well to Merezhkovsky.

I enrolled as a graduate student in the department of genetics at the University of California at Berkeley in 1960. Although I was twenty-two years old and already a mother of two persistently active boys, my enthusiasm for pursuing cell genetics and evolution overwhelmed any thoughts of becoming a full-time housewife. More than husbands, I had always wanted children. Unlike my own parents, I found whisky and cigarettes, poker and bridge, meetings and politics, gossip and golf, unbearably boring. I, on the other hand, was insufferably bookish, serious, and studious and preferred the company of babies, mud, trees, fossils, puppies, and microbes to the normal world of adults. I still do.

At Berkeley there was absolutely no relationship between members of the department of paleontology, where evolution was studied, and those of the department of genetics, where evolution was barely mentioned. Because I sought an education in all aspects of evolution, paleontology, and genetics that could illuminate the evolutionary history of cells, at first I was shocked by the depth of academic apartheid. Each department seemed oblivious of people and subject matter beyond its borders. Furthermore, since nearly all the bacterial geneticists at the Bacterial-Virus Laboratory (BVL), on the east side of the campus, began their careers as chemists, most were abysmally uninformed about the genetics of plant and animal cells. Few had ever heard of cytoplasmic heredity or of organelles in cells with nuclei. No bacterial geneticist, microbiologist, or virologist at that end of campus knew anything about genetic systems in the cytoplasm of algae. Some were so bacteriocentric that they barely understood mitosis, the kind of cell division characteristic of cells with nuclei. Certainly they never taught or even thought about the special variation on mitosis, called meiosis, that underlies the rules of Mendelian heredity. Many knew almost nothing about that heredity, which applies only to nucleated organisms. So arrogant were these johnny-

come-lately biologists—trained in physics and chemistry—
that they did not even know they did not know. Many transmitted chemical sophistication but biological ignorance and
arrogance to their graduate students. The BVL faculty and students had not even heard of the exciting work in cytoplasmic
inheritance in a then-thriving branch of genetics, ciliate genetics. Even members of the genetics department, on the west
end of campus, were unaware of the ciliate genetic science
that was so interesting to me. Their lack of interest and ignorance surprised but did not deter me. I had been fascinated by
the genetics of *Paramecium*, the ciliate, and by its leader,
Tracy Sonneborn (1895–1970), since I had first read about it.
What Sonneborn and his French colleague Jannine Beisson
had discovered seemed grossly to contradict the ubiquitous
dogma that induced characteristics cannot be inherited. Sonneborn, longtime professor of genetics at Indiana University,
with researcher Beisson reported that if *Paramecium* cilia are
surgically removed in clumps with their bases and turned
around some 180 degrees on the cell's surface, then replaced,
the cilia will appear in offspring cells, for many future generations, in this reversed position. In other words, the cilia reproduce and the change experimentally induced by the scientists
was inherited, for at least two hundred generations after the
operation. Here was a laboratory example of the so-called
inheritance of acquired characteristics that orthodoxy dismissed as Lamarckianism.

Pursuing my interest in such details was even then a
lonely intellectual exercise. As the political orientation of
the 1960s accelerated, more time and more talk in academia
was dedicated to "relevance," with the result that any intellectual pursuit was assessed in relation to human welfare. In
this climate my interest in the patterns of cell inheritance
was antisocial. What preoccupied me most was irrelevant to
my instructors and most of my fellow students.

Genetics, despite the Berkeley bacterial geneticists, still
seemed to me to be the key to evolutionary history. I col-

lected more examples of non-Mendelian (nonnuclear) heredity in a diverse array of species: plants such as *Eupatorium*, *Zea* (corn), *Mirabilis jalapa,* and *Oenothera*, and algae such as *Chlamydomonas*. I studied the nonnuclear oxygen-respiration-deficient mutants of yeast called "petites," which grow slowly and form small colonies. I also reviewed the "kappa-killer" inheritance pattern of the ciliate *Paramecium*. This beautifully researched phenomenon was described by Tracy Sonneborn, who saw that some paramecia were born to kill others who were genetically distinct. And I never felt that "nonnuclear" heredity was "unclear." H. J. Muller (1890–1967), a geneticist who later received the Nobel prize for showing how X-rays induce hereditary change (mutation), had proclaimed the existence of naked genes, at least in principle, at the core of life. Despite his elegant work, I did not believe anyone had proved that "naked genes" exist beyond the nucleus of plant or animal cells. I pored over the old but brilliant work of Edouard Chatton (1883–1947), a French marine biologist, and Lemuel Roscoe Cleveland (1892–1969), a Harvard professor and research scientist. I read myriad papers by Tracy Sonneborn, who wrote beautifully and thought aloud as he worked. I pursued in print the gorgeous high-power photographs of organelles called electron micrographs that had been taken in Madison by my iconoclastic and dedicated University of Wisconsin professor Hans Ris. These disparate sources of information substantiated my hunch. Bacteria, not naked genes, did reside outside the nucleus but inside the cells of certain protists, yeasts, and even plants and animals. As I reviewed the literature on cytoplasmic genetics it became obvious that at least three classes of membrane-bounded organelles (plastids, mitochondria, cilia), all outside the nucleus, resembled bacteria in their behavior and metabolism. Indeed, differentiating between a bacterium trapped in a cell and an organelle inherited as part of a cell seemed to me, in some cases, impossible. A trapped blue-green bacterium that shed its wall to reside and grow comfortably in the cytoplasm of a plant cell seemed to be exactly the organelle everyone called a chloroplast.

Emboldened by intellectual forays into this literature of the genetics of cytoplasmic organelles, I predicted that the plastid, née trapped bacterium, must have retained some bacterial DNA.

What began as library sleuthing in the genetics department at UC Berkeley continues to this day. I still greedily acquire scientific papers on microbial symbionts and membrane-bounded organelles. My students and I still work on the central idea: the origin of cells with nuclei is exactly equal to the evolutionary integration of symbiotic bacterial communities.

My earliest complete statement of "serial endosymbiosis theory" was published after fifteen or so assorted rejections and losses of an early, painfully convoluted, and poorly written manuscript. Called "Origin of Mitosing Cells," it was finally accepted for publication in 1966 through the personal intervention of James F. Danielli, then editor of the daring *Journal of Theoretical Biology*. Of course, the article, when it finally appeared in print in late 1967, carried my first married name, Lynn Sagan. The theory was dubbed SET, the acronym for *Serial Endosymbiosis Theory* (not to be confused with SETI—the search for extraterrestrial intelligence), by another protist aficionado, Professor Max Taylor of the University of British Columbia, Vancouver.

Not until I was well into my second marriage in 1969 and pregnant with my daughter Jennifer was I obliged to stay home for extended periods. Enforced home leave permitted uninterrupted thought. This, in turn, stimulated me to document the expanded version of my four-part SET narrative clearly. The story of the origin of cells begun in my 1967 paper sprouted, expanded, and eventually was pruned into a book-length manuscript. I typed late into many nights, determined to make the deadline required by contract. Of course, as a virtual unknown I was given neither advance nor compensation for the many illustrations I commissioned. All help came from home. Finally I completed what I thought was the final draft. With pride and care, even earlier in the morning than the voices of children, I boxed up and then mailed off the heavily illustrated work to the publisher who held the contract: Academic Press in New York City. The receipt of the box was not acknowledged. I waited. I continued to wait, for about five months. One day my box, without explanation, sent by surface book rate, reappeared at my mailbox. Much later I was informed, not even by the editor, that extremely negative peer review had led Academic Press to hold the manuscript for months. From the press finally I received a form letter of rejection. No explanation, in fact not even a personal letter signed by the Academic Press editor, accompanied the formal rejection. More than a year later, after far more painful and far longer labor than Jenny ever caused me, the book finally was nicely edited, produced, and published by Yale University Press. Because of commentary and criticism by Max Taylor and other generous colleagues the serial endosymbiosis theory prevailed and eventually the pain of the Academic Press rejection subsided.

SET attracted experimental contributions by many scientists and graduate students unknown to me throughout the 1970s and 1980s. Molecular biological, genetic, and high-powered microscopic studies all tended to confirm the once-radical nineteenth-century idea that the cells of plants and of our animal bodies, as well as those of fungi and all other organisms composed of cells with nuclei, originated through mergers of different types of bacteria in a specific sequence. Joint residence prevailed and proliferated. My most current version of SET is shown in Figure 2. Today I am amazed to see a watered-down version of SET taught as revealed truth in high school and college texts. I find, to my dismay if not to my surprise, that the exposition is dogmatic, misleading, not logically argued, and often frankly incorrect. Unlike the science itself, SET now is uncritically accepted. So it goes.

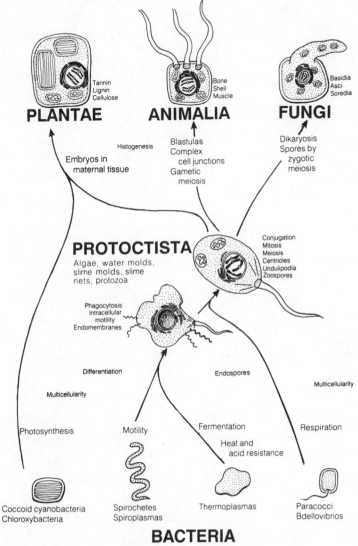

FIGURE 2

SET (serial endosymbiosis theory) phylogeny

SET is a theory of coming together, of merging of cells of different histories and abilities. Before SET there was no cell-fusion sex like that of the egg fertilized by the sperm. SET made our kind of fusion sex possible. Sex, too, is the coming together, the merging of cells of different histories and abilities.

....

CHAPTER 3

INDIVIDUALITY BY INCORPORATION

Low at my problem bending,
Another problem comes —
Larger than mine — Serener —
Involving statelier sums (69)

Symbiosis, the term coined by the German botanist Anton deBary in 1873, is the living together of very different kinds of organisms; deBary actually defined it as "living together of differently named organisms." In certain cases cohabitation, long-term living, results in symbiogenesis: the appearance of new bodies, new organs, new species. In short, I believe that most evolutionary novelty arose and still arises directly from symbiosis, even though this is not the popular idea of the basis of evolutionary change in most textbooks.

My theory of the symbiogenetic origin of plant, animal, and other cells with nuclei employs four provable postulates. All four involve symbiogenesis, incorporation, and body fusion by symbiosis. The theory precisely outlines the steps that must have occurred in the past, especially in relation to the bright green cells of plants. Cells, of course, are familiar units of structure in mosses, ferns, and all other plants. The slender stamen hairs particularly visible in *Zebrina* and *Tradescantia* ("wandering jew") flowers are made

of rows of such plant cells. Large, walled green cells preceded plants: they were already fully formed in the green
algae, water-dwelling ancestors of plants. That nucleated
organisms evolved by merger is best appreciated in plants
because in their large and beautiful cells, the integrity of
their component organelles is easily observed. The idea is
straightforward: four once entirely independent and physically separate ancestors merged in a specific order to become
the green algal cell. All four were bacteria. Each of the four
bacteria types differed in ways we can still infer. In both
merged and free-living forms the descendants of all four
kinds of bacteria still live today. Some say the four types are
mutually enslaved, trapped both in the plant and as the
plant. Today each of the types of former bacteria provides
clues about its ancestry; life is chemically so conservative
that we can even deduce the specific order in which they
merged. The term *serial* in *serial endosymbiosis theory* refers
to the order in the merger sequence.

I believe I have now convinced many scientists and students that parts of cells, the organelles, originated symbiogenetically, as a consequence of different permanent symbioses. Of course, very little evidence from the theory is my
work; hundreds of scientists contributed. I'm now working
on expanding the idea to show that other organisms larger
than cells and their new organs and new organ systems also
evolved by symbiogenesis.

If symbionts merge entirely, if they fuse and form a new
kind of being, the new "individual," the result of the merger,
by definition, evolved through symbiogenesis. Although the
concept of symbiogenesis was proposed a century ago, only
now do we have the tools to test the theory rigorously.

I'll try to outline the idea as simply as possible. First, a
sulfur- and heat-loving kind of bacterium, called a fermenting "archaebacterium (or "thermoacidophil"), merged with a
swimming bacterium. Together the two components of the

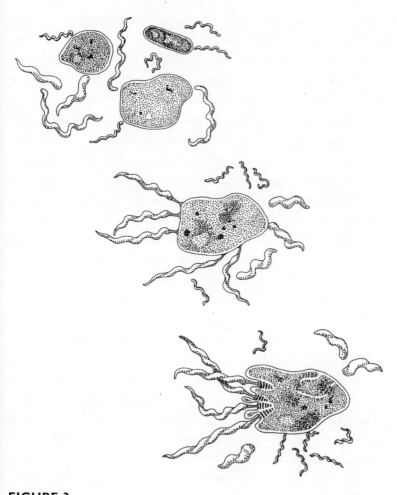

FIGURE 3

Spirochetes become undulipodia

integrated merger became the nucleocytoplasm, the basic substance of the ancestors of animal, plant, and fungal cells. This earliest swimming protist was, like its descendants today, an anaerobe. Poisoned by oxygen, it lived in organic rich muds and sand, in rock crevices, puddles, and pools where oxygen was absent or scarce. (Fungal cells include those of mushrooms and yeast). Animal, plant, and fungal cells are all nucleated cells because, watery and translucent, they contain a visible nucleus. In plants and animals the membrane-bounded nucleus disappears, as the membrane dissolves and the chromosomes become visible each time a cell multiplies by division. The chromatin, the red-staining material from which the chromosomes are made, coils into easily seen structures. The textbook explanation of this process is that chromatin condenses into visible, countable chromosomes whose number is representative of the species in question. The dance of the chromosomes, which then disappear into loosely coiled chromatin as the nuclear membrane reappears constitutes steps in the cell division process of mitosis. Mitosis, with its many variations in protoctistan and fungal cells with nuclei, evolved in the earliest organisms with nuclei.. After mitosis in swimming protists evolved another type of free-living microbe, an oxygen-breathing bacterium, was incorporated into the merger. Even larger, more complex cells arose. The oxygen-breathing three-way complex (acid heat lover, swimmer, and oxygen breather) became capable of engulfing particulate food. Complex and startling beings, these cells with nuclei, swimming and breathing oxygen, first appeared on Earth perhaps as early as some 2,000 million years ago.

This second merger, in which the swimming anaerobe acquired the oxygen-breather, led to three-component cells increasingly capable of coping with accumulating levels of free oxygen in the air. United, the subtle swimmer, the acid- and heat-tolerant archaebacterium, and the oxygen breather now formed a single individual and prolifically generated clouds of offspring.

In the final acquisition of the complex-cell-generating series, oxygen breathers engulfed, ingested, but failed to digest bright green photosynthetic bacteria. The literal "incorporation" occurred only after a great struggle in which the undigested green bacteria survived and the entire merger prevailed. Eventually the green bacteria became chloro-plasts. As the fourth partner, these productive sun lovers became entirely integrated with the other formerly separate partners. This final merger gave rise to swimming green algae. Not only were the ancient swimming green algae ancestors of today's plant cells, but all their individual com-ponents are currently alive and well, still swimming, fer-menting and breathing oxygen.

My best work, I believe, is the development of the details of the serial endosymbiosis theory. The central idea is that extra genes in the cytoplasm of animal, plant, and other nucleated cells are not "naked genes": rather they originated as bacterial genes. The genes are a palpable legacy of a vio-lent, competitive, and truce-forming past. Bacteria, long ago, which were partially devoured and trapped inside the bodies of others, became organelles. Green bacteria that photosyn-thesize and produce oxygen, called *cyanobacteria*, still exist in ponds and streams, in muds, and on beaches. Their rela-tives cohabit with countless larger organisms: all plants and all algae.

The reason the early plant geneticists discovered genes in the chloroplasts of plant cells is that they are always there. Little green descendants of cyanobacteria are in every plant cell at all times.

Cyanobacteria are a wildly successful form of life. They coat our shower curtains and form scums on our swimming pools, toilets, and ponds. If warmed and sunlit, they can color a standing puddle bright green in a few days. Although most cyanobacteria are still free-living, some live with very different partners as symbionts. Some are chloroplasts, the green parts of algal and plant cells. Others live in leaf cavi-ties, root layers, or stem glands of green plants.

Just as cyanobacteria and chloroplasts are close relatives, so are mitochondria related to free-living oxygen-breathing bacteria. The lineal ancestors of animal and plant mitochondria, I claim, as did other earlier, often ignored scholars, also began as free-living bacteria. Mitochondria, intracellular power factories, produce chemical energy inside the cells of all animals—and plants and fungi. Mitochondria are also regular residents in most of the myriad obscure microbial beings, the protoctists, from which plants, animals, and fungi evolved. By sheer numbers, chloroplasts and mitochondria, rather than humans, are Earth's dominant life-forms. Wherever we humans go, the mitochondria go too, since they are inside us, powering all our metabolism: that of our muscles, our digestion, and our thoughtful brains.

Symbiogenesis, an idea proposed by its Russian inventor Konstantin Merezhkovsky (1855–1921), refers to the formation of new organs and organisms through symbiotic mergers. As I will show it is a fundamental fact of evolution. All organisms large enough for us to see are composed of once-independent microbes, teamed up to become larger wholes. As they merged, many lost what we in retrospect recognize as their former individuality.

I am fond of bragging that we, my students and colleagues, have won three of the four battles of serial endosymbiosis theory (SET). We can now identify three of the four partners in the origin of cell individuality. Scientists preoccupied with this story now agree that the ground substance of cells, the nucleocytoplasm, descended from archaebacteria; in particular, most of the protein-making metabolism arises from thermoacidophil ("thermoplasmalike") bacteria (step 1). The oxygen-respiring mitochondria in our cells and other nucleated cells evolved from bacterial symbionts now called "purple bacteria" or "proteobacteria" (step 3). The chloroplasts and other plastids of algae and plants were once free-living photosynthetic cyanobacteria (step 4). Note that step 2 is not described.

One major, contentious issue remains: how did the swimming appendages, the cilia, originate? Here is where most scientists part company with me. They agree with Max Taylor's description of my version of symbiotic theory as "extreme SET." Taylor and his colleague, Tom Cavalier-Smith, at the University of British Columbia in Vancouver favor a nonsymbiotic, "branching" theory of the origin of the earliest nucleated cells. Theirs is still the prevalent hypothesis. But evidence exists that an enigmatic second bacterial partner joined the ancient alliance. The first merger, the permanent fusion of the first and second partners, was crucial. It happened. Even if the legacy of the first merger is obscure and difficult to detect today, it left clues, and we seek these clues. My hypothesis that the swimmer, another different microbe, was symbiotically acquired in the very first, most ancient step in the origin of nucleated cells is the least defensible component of the idea. This first merger occurred perhaps 2,000 million years ago. The key idea (step 2) of SET is that cilia, sperm tails, sensory protrusions, and many other appendages of nucleated cells arose in the original fusion of the archaebacterium with the swimming bacterium. I predict that within a decade we will win this argument: eventually we will be four for four! I explain why I hold my unpopular opinion in this chapter and give you an idea of why I spend my life collecting evidence from all dusty corners of biology. Some colleagues label me combatative; others, unfair. Some say I only collect relevant work and unfairly ignore contradictory data. These accusations may be correct.

Bacteria, merging in symbiosis, leave us hints of their former independence. Both mitochondria and plastids are bacterial in size and shape. Most importantly, these organelles reproduce so that many are present at one time in the cytoplasm but never inside the nucleus. Both types of organelles, plastids and mitochondria, not only proliferate inside cells but reproduce differently and at different times from the rest of the cell in which they reside. Both types, probably 1,000 million years after their initial merger, retain their own

depleted stores of DNA. The ribosomal deoxyribonucleic acid (DNA) genes of mitochondria still strikingly resemble those of oxygen-respiring bacteria living on their own today. The ribosomal genes of plastids are very much like those of cyanobacteria. In the early 1970s, when sequences of nucleotides in the DNA of plastids in algal cells were first compared with sequences from free-living cyanobacteria, chloroplast DNA was found to be far more similar to cyanobacteria DNA than it was to the DNA in the nucleus of the same algal cell! The case was closed. The three-way comparison of these DNA sequences (from nucleus, from organelle, and from the organelle's free-living codescendant) proved the bacterial origins of plastids. An analogous case was made for the less colorful organelles, the mitochondria. No time-travel witness was needed to tell the tale.

Max Taylor is the world's greatest expert on certain colorful and fascinating marine protists. He and nearly everyone else calls them dinoflagellates. I stubbornly refuse to conform. I call them *dinomastigotes* because I'm allergic to the term flagella when it is applied to nonbacteria. Bacteria bear flagella and organisms with a nucleus never do, in my view.

At one point Max explicitly developed alternative hypotheses to my serial endosymbiosis theory. In the early 1970s, he outlined an endogenous or "direct filiation," theory. He made explicit the outline of a nonsymbiotic theory for the origin of nucleated cells. His was directly opposed to my view. Direct filiation theory holds that the three kinds of cytoplasmic organelles—mitochondria, cilia, and chloroplastids evolved without symbiosis. Rather, by Max's hypothesis and all presymbiogenetic opinion, they all arose by "pinching off" DNA from the nucleus. Mitochondria, plastids, and cilia always were parts of cells. They never began as strange bacteria. Not only was direct filiation Max's theoretical baby, but it accorded with the unstated assumption of branching, versus fusing evolution held by all biologists.

To be thorough and to organize vast amounts of detailed information, Max catalogued alternative possible versions of direct filiation and symbiosis theory. Direct filiation, in its extreme version, denies the existence of any cell symbiosis. Nucleated cells evolved directly from separate origins or from changes in a single type of bacterial ancestor. Plastids, the generic term for photosynthetic organelles in algae and plants, include the famous chloroplasts of green algae and their plant descendants. The category also includes red rhodoplasts of red seaweeds and brown phaeoplasts found in dinomastigotes, diatoms, phaeophytes, and many other algae. The mildest version of symbiosis theory that posits only plastids evolved from symbiotic photosynthetic bacteria and that mitochondria and all other organelles arose without symbiosis, by direct filiation from genes that left the cell nucleus. Then there is a middle-ground SET that accepts symbiotic origins for both plastids and mitochondria. The middle-ground version is now indisputable. As explained in textbooks, a preponderance of the evidence favors the idea.

Max Taylor is not being unfair when he labels me a radical symbiogeneticist and dubs my version of SET "extreme." Why? In spite of slim evidence, I still believe in the swimmer of Figures 2 and 3. The other class of organelles (step 2 in the scheme) is also bacterial in origin, in my view. Cilia, sperm tails, sensory protusions, and other cell whips, always underlain by tiny dots, bodies called *centriole-kinetosomes*, come from step 2. These structures are associated with cell movement. The organelles in step 2, since they are even more ancient and tightly integrated into cells than are either mitochondria or plastids, enjoy an evolutionary history most difficult to trace.

Part of the problem in explaining the origin of organelles of step 2 is that this class of moving cell structures has many names, all confusing. The crucial part of the theory begins with the cilium's underlying dot.

The enigmatic centriole-kinetosomes act as tiny seeds. Sperm tails, cilia, and in some organisms, the mitotic spindle crucial to moving chromosomes at cell division, extend from the little seeds almost magically. A cell, depending on its parentage, may grow one or many centriole-kinetosomes, either in connection with preexisting centrioles or apparently from nowhere (the latter is termed *de novo* appearance of centriole-kinetosomes). Timing is crucial. A naked cell can form many centriole-kinetosome "seeds" that all make protrusions at once. All centriole-kinetosome "seeds" are made of thin tubes of protein, the microtubules. The protein of the walls of the tubes, naturally enough, is called tubulin.

Acceptance of symbiotic origin for mitochondria and plastids was finalized with the discovery that both these kinds of organelles contain distinct DNA, separate from that of the nucleus and unequivocally bacterial in style and organization. The DNA of these organelles codes for their own peculiar proteins. Just as in free-living bacteria, protein synthesis occurs inside mitochondria and plastids. Ford Doolittle and Michael Gray, molecular biologists at Dalhousie University in Halifax, Nova Scotia, show DNA sequences in mitochondria and plastids extremely similar to those in selected bacteria living on their own. At this point the scientists accept this and much other evidence as proof of the moderate, the only-three-out-of-four, version of SET.

But what about my "extreme" theory? Is there a different bacterial remnant, a centriole-kinetosome ancestor of eukaryotic nucleated cells? I think the integration of the centriole-kinetosome bacterium is what made the eukaryotic cell in the first place! If I am correct, symbiogenesis is the factor that distinguishes all nucleated-cell life from all bacterial life. No middle ground exists—either a group of organisms evolved by symbiogenesis or it did not. My claim is that all nucleated organisms (protoctists, plants, fungi, and animals) arose by symbiogenesis when archaebacteria fused with ancestors of centriole-kinetosomes in the evolution of the ultimate protoctist ancestor: the nucleated cell.

The former stranger that became the centriole-kinetosome still has free-living relatives. They are the bacteria known as spirochetes. Their ancestors, ancient wild wrigglers, hungry and desperate, invaded many archeabacteria, even some similar to today's *Thermoplasma*. Invasions were followed by truces. I postulate that the earliest nucleated cells emerged after spirochetes and archeabacteria survived in the merged state. The nucleated cell evolved by symbiogenesis.

Other "extreme" versions of SET exist. Without developing details, Dr. Hyman Hartman, a researcher working at the NASA Ames Research Center in Moffett Field, California, suggests the nucleus itself began as a free-living bacterium. The "nucleus as symbiont" view was first articulated not by bushy-headed Hartman, who is very much alive, but by the Russian K. S. Merezhkovsky who died in 1921. Merezhkovsky never said mitochondria began as symbionts because when he wrote neither he nor anyone else understood what mitochondria were. Until the advent of the electron microscope, a very high-power instrument, there were over twenty terms for the little bodies inside cells, later recognized as mitochondria. The terms were consolidated and their meaning made clear in the midsixties as microscopy improved.

I do not agree with Hartman or Mereszhkovsky that nuclei have symbiotic origins. The microbial world does not, as far as I know, contain free-living nucleuslike bacteria. The nucleus, in my view, evolved in response to the uneasy merger of *Thermoplasma*-like and *Spirochoeta*-like bacteria. "New cells" emerged, they enlarged, and their interacting membranes proliferated. Their genetics became more complex because of their double ancestry.

Radney Gupta, of McMaster University, continues to present detailed, if arcane, evidence for the "chimeric" nature of the earliest nucleated cells. Gupta's arguments are based exclusively on analysis of the sequence of amino acids in many indispensable proteins. He uses different terms and separate criteria but our basic ideas are the same: archaebacterial-eubacterial fusion spawned the first cell ancestors with membrane-bounded nuclei.

The origin of nucleated cells, everyone agrees, was a crucial innovation in the evolution of life on Earth. The first microbes to sport nuclei were small oxygen-avoiding swimmers. Today they would be classified in the protoctist kingdom. The smallest living members of this motley group are as tiny as bacteria. Even though they live in zones where oxygen is excluded, because they have nuclei and many other traits of nucleated cells, they are not bacteria.

What happened? When compared to any extracellular environment, always exposed to desiccation, food depletion, poisoning, and other potential tragedy, any intracellular milieu is watery and well nutrified. Any spirochete (or other swimming bacterium) that penetrated the membrane barrier of an archaebacterium would enjoy a constant stream of energy and food. The reproduction rates of the attacker and the attacked eventually became coordinated. Swimmer-attacker survivors who overwhelmed their living home would not survive for long. We know today that attackers become symbionts and, with time, eventually can become organelles. New survival tricks emerge after mergers. I envisage wriggling, oxygen-poisoned cells attached to the edges and insides of archaebacteria, as always in search of regular meals. Inhabited by wrigglers, the infected archaebacteria increased their speed because their foreign attachés never stopped moving. Nucleated cells divide to reproduce by mitosis, a process dubbed the "dance of the chromosomes." Elsewhere I have described how I think this kind of cell division originated from the incessant movement of live spirochetes.

My former and current graduate students and I continue to test the "extreme" hypothesis. In our hypothetical scenario, remnants of the ancient merger are detectable in the modern behavior and chemical reactions of all nucleated cells. Of course we need more evidence even to convince ourselves fully.

The order of events in evolution is decisive. Corkscrew-shaped spirochetes, speedsters of the microbial world, are coiled and snakelike. Through viscous fluids such as mud, slime, mucus, and living tissue, these bacteria dart to and fro, up and down, left and right. As they do now, in the remote past they outswam other bacteria. Quick and prolific, spirochetes invaded archaebacterial inner space, and those that interacted survived. Their living descendants are now inextricably involved in mitotic movements and other actions of complex cells. The partners are so fused that the origin scenario is difficult to reconstruct—but not impossible.

All cells with mitochondria also have microtubules, remnants of the ancient wrigglers. Such structure is consistent with the idea that the spirochete-archaebacteria symbiosis was established first. Today certain mitotic swimming cells, to which oxygen is poison, still lack mitochondria. I deduce that the mitotic ancestor to all of eukaryotic life evolved before oxygen permeated all corners of the atmosphere.

Spirochetes today swim in both oxygen-rich and oxygen-poor environments. They sometimes attach to neighbors so skillfully that biologists mistake their attachment points for centriole-kinetosomes and their bodies for cilia. Today's spirochetes dwell in vast numbers in the intestines of healthy wood-eating insects. A few kinds inhabit human intestinal or testicular tissue. Some thrive in mud. Others dwell on the leaky membranes of protists such as mud ciliates or trichomonads. Spirochetes, in general, thrive in wet, rich, dark habitats. The life of the spirochete is to wriggle, feed, and clone. They reproduce by transverse bacterial-style division. The road to cilia began when the early spirochetes moved inside vulnerable neighbors wherever they were able and some never returned to the outside. From the mutual motion of many small spirochetes, after much integration, nucleated swimmers, the first protists, evolved.

I still hope that the final SET postulate will prevail. Many colleagues have told me to give up.

Centrioles and kinetosomes, like a friendly Dr. Jekyll and Mr. Hyde, are never both seen in dividing cells at the same time. In many cells, centrioles transform to kinetosomes that grow their shafts as soon as mitotic cell division finishes. This points to a single identity. In 1898 L. F. Henneguy, physiology professor at Paris, and Mihaly von Lenhossek, in Budapest, noted the identity of the centriole with the kinetosome in animal tissue and wrote about it. Their idea that the mitotic centrioles reproduce, move from the mitotic poles, and become the kinetosomes of cilia is the "Henneguy-Lenhossek theory." Proved by electron microscopy after the deaths of both scientists, the validity of the Henneguy-Lenhossek theory inspires my use of the double name centriole-kinetosome. Spirochetes, I believe, originally formed attachment structures to susceptible archaebacteria. As they symbiogenetically integrated the point of their attachment became the centriole-kinetosome of today. The Oxford University biologist David C. Smith compares the theorized remains of symbiotic spirochetes in the innards of nucleated cells to Lewis Carroll's Cheshire Cat. Just as the cat slowly disappears but leaves a grin enigmatically floating in midair, "the organism progressively loses pieces of itself, slowly blending into the general background, its former existence betrayed by some relic."[1] The merged being becomes something inside the participating partner. As the fusion is complete, it is difficult to determine the relative genetic contributions of the partners.

In our lab we search for nucleic acids and proteins more common among free-living spirochete bacteria and cilia than among other arbitrarily chosen organisms or cell organelles. Many studies are under way, mostly at medically oriented laboratories whose scientists are unconcerned with evolution. Mainly I just monitor the findings.

Some cells can withstand freezing to temperatures near absolute zero. Metabolism ceases. The flow of food, waste, and energy stops. Yet when rewarmed, these cells function and grow perfectly. The cell remembers; the information of life is intrinsic to its cellular structure. Severed sperm tails, without nuclei, mitochondria, and even outer membranes, survive and swim for as much as an hour in the proper balanced and energized solutions. These sperm tails, the cilia of ciliate protists, the cilia on the cells of women's fallopian tubes, and the cilia in our throats (all forms of undulipodia with the characteristic nine sets of microtubules of their detailed structure), I think, are derived from originally free-living spirochetes that integrated into our ancestral archaebacteria. I am optimistic that we will eventually have proof, when the genes for the motility proteins and—even more important, because they change less rapidly—the motility and other relevant proteins themselves—are identified and sequenced.

The most crucial datum for the bacterial origin of centriole-kinetosomes is the discovery of what could be remnant DNA in a green swimming microbe. John Hall, David Luck, and Zenta Ramanis of Rockefeller University have found, in the green alga *Chlamydomonas*, special genes that coded for traits that influence centriole-kinetosome-microtubule swimming structures. These genes were grouped together and separate from standard nuclear genes. As soon as I read the work of the Rockefeller scientists I felt convinced of the validity of "extreme" SET. No one has yet separated centriole-kinetosome DNA from other cell DNA, so, of course, centriole-kinetosome genes have not been directly compared to genes claim free-living spirochetes. Furthermore, these scientists claim the special set of genes related to the centriole-kinetosomes and their undulipodia are within the nucleus. They have photographed DNA very near the alga's two reproducing centriole-kinetosomes. The centriole-kinetosome DNA, distinguishable from the rest of the nuclear DNA in some stages of the cell's development, joins all the other chromosomal DNA during mitotic cell division.

Unable to detect the centriole-kinetosomal DNA in this same green alga, Joel Rosenbaum, at Yale University, and other researchers deny its existence. Even circumstantial evidence for the spirochete hypothesis is tantalizingly scant. I am prepared to be incorrect. Perhaps nonspirochete bacteria, such as a green nonsulfur type of eubacterium as Gupta hypothesizes, long ago infused other cells. Perhaps symbiosis was not even involved in the origin of the very earliest swimming eukaryotes. Certainly Carl Woese and Max Taylor's colleague Cavalier-Smith both disagree with me. They see no role of symbiogenesis at all in the origin of the earliest anaerobic protists. Nevertheless, I believe it does not take Sherlock Holmes to suggest that in the ancient past a microbial actor greatly increased the swimming speed of what became a consortium. The identical descendant of the motility remains.

The neurons, nerve cells in our brains, and peripheral nerves are replete with microtubules made of tubulin protein. The same microtubules, exactly the same, make up the cilia, sperm tails, and walls of the centriole-kinetosomes. The axons and dendrites, extensions of nerve cells by which we process information in our brains, are underlain by microtubules. Our brains themselves and the thought required to read this sentence, if my radical symbiogenesis theory is correct, were made possible by the protein microtubules that first evolved in bacteria. Even if research shows

my spirochete hypothesis to be untrue, thinking about symbiosis is itself a symbiotic phenomenon. The oxygen we
breathe enters the brain from our bloodstream and is incessantly metabolized by the mitochondria that we know are
former respiring bacteria. Whether or not spirochete wrigglers are at the core of our being, we remain symbiotic beings
on a symbiotic planet.

CHAPTER 4

THE NAME OF THE VINE

Much Madness is divinest Sense —
To a discerning Eye —
Much Sense — the starkest Madness —
'Tis the Majority
In this, as All, prevail —
Assent — and you are sane —
Demur — you're straightway dangerous —
And handled with a Chain — (435)

Names of live beings seem harmless enough. So do groupings. Yet the superficially boring practice of naming and grouping has profoundly affected my scientific life. Faulty taxonomics misleads with the subtlety of unstated assumptions or religious beliefs. That symbiogenesis collides directly with common, cherished presumptions is one reason its acceptance has been delayed.

Taxonomy is the science of identifying, naming, and classifying organisms. Names and classification schemes organize great quantities of information. Taxonomies, like maps, bring into relief selected distinguishing features. However, in the phrase popularized by the English-American philosopher-anthropologist Gregory Bateson, "The map is not the territory." Nor is the name the organism. The history of any organism is often depicted on a family tree. Family trees usually are grown from the ground up: a single trunk branches off into many separate lineages, each branch diverging from common ancestors. But symbiosis shows us that such trees are idealized representations of the past. In reality the tree of life often grows in on itself. Species come together, fuse, and make new beings, who start again. Biologists call the coming together of branches—whether blood vessels, roots, or fungal threads—anastomosis. Anastomosis, branches forming sets, is a wonderfully onomatopoetic word. One can hear the fusing. The tree of life is a twisted, tangled, pulsing entity with roots and branches meeting underground and in midair to form eccentric new fruits and hybrids. Anastomosis, although less frequent, is as important as branching. Symbiosis, like sex, brings previously evolved beings together into new partnerships. Like sex, too, some symbioses are protracted unions with stable, prolific futures. Others quickly dissolve. The interaction of each generation of genetically continuous beings calls into question any picture-book tree of life.

Using the now-famous drawing worked on by many and created in the mid-1980s by Dorion Sagan, my eldest son, we employ the metaphor of a hand (Figure 4). Each of five fingers represents a major group of organisms. Think of each digit as one of five great realms: all the bacteria (the monerans or prokaryotes, whose cells lack nuclei), the protoctists (all the algae, slime molds, ciliates, and many other obscure organisms formed through symbiogenesis and composed of cells with nuclei), the animals (all of which develop from embryos that develop from sperm-egg unions), the fungi (yeasts, mushrooms, and molds that grow from fungal spores), and the plants (which develop from both spores and, at other times, sexually produced embryos even though all do not photosynthesize). Except for bacteria, each finger on the hand of life has multiple symbiotic microbial ancestors.

FIGURE 4

Five-kingdom hand

When Ivan Wallin, at Columbia University in the 1920s, proposed that cell components, chloroplasts and mitochondria, originated as symbiotic bacteria, he and his ideas were roundly rejected. In polite biological society, his "symbionticism theory" was ridiculed. He was ostracized by the serious research community and, at age forty, gave up his laboratory investigations of symbiosis. Bacteria were agents of disease, not originators of evolutionary novelty. Microbes, his colleagues insisted, did not jump from their lives as free-living bacteria to become captive parts of animal cells. Wallin moved from New York to Denver, where his career as a teacher-scholar in the university thrived for another four decades, but he never again argued about symbiotic origins of organelles. Bacteria were dangerous pests; no one even wrote about them in the context of life's evolution. Furthermore, the limbs on family trees branched. No one, except Merezhkovsky, argued that branches fused.

Now we know that the work of Wallin, the "eccentric," is more in tune with the thinking of our time than with that of his own. One reason his claims were not even critically examined was widespread commitment to a rigid taxonomy. All organisms perceived as either animal or plant were classified accordingly. So be it. Indeed, confusingly but revealingly, microbes that swam were assigned to the animal kingdom. When extremely similar microbes were green and temporarily nonmotile, they were declared to be plants. A botanist, just because she dwelled in a botany department, would classify all the microbes and their descendants as plants. A zoologist across campus would assign a very similar being to his animal kingdom. Contradictions proliferated. Difficult-to-see small organisms, even identical ones, were claimed by botanists as plants and by zoologists as animals. Microbiologists (bacteriologists) and mycologists (mold and mushroom experts) fought similar battles over bacteria, yeast, and other fungi. Regretably, this picturesque taxonomic confusion still abounds.

We can excuse Antony van Leeuwenhoek (1632–1723), inventor of the microscope, from calling the new creatures he discovered animalcules. With what else was he to compare them? But I believe it is inexcusable for scientists today to retain the term *protozoa* (which translates as "first animals") for swimming creatures whose biological characteristics are definitely not those of an animal. Some former protozoa, microbes with nuclei, although ancestral to animals, are also ancestral to plants, to fungi, and to all the unruly protoctists. The ancestral protists themselves, which never develop from sperm or egg or animal embryo, are not animals. The group is shockingly diverse. More than fifty major lineages live on.

Included among them are diatoms, brown algae, ciliates, and many less well known groups of beings representing the "everywhereness of life."[1] Whether ameba, ciliate, hypermastigote, or other, they are not animals.

I wince too when I hear biologists use the phrase "blue-green algae." No such things exist; these blue-green wonders are in every way photosynthetic bacteria. Equally irritating is "one-celled animal," also nonexistent. Nor do "higher plants" or "multicellular plants" exist. All animals and all plants develop from embryos, which are by definition multicellular. Since all plants and all animals are multicellular, the adjective is redundant. The characterization as a "protozoon" is an oxymoron and the wordy phrases "multicellular plant" and "multicellular animal" are misleading.

Language can confuse and deceive. These antiquated terms—"blue-green algae," "protozoa," "higher animals," "lower plants," and many others—remain in use despite their penchant to propagate biological malaise and ignorance. The use of these insults to the living benefits those people whose budgets, class notes, and social organization depend on their continuity. I suggest that one reason Wallin's good ideas were opposed or ignored was that he was thoroughly misunderstood by the many biologists and teachers who reinforce the misconception of fixed classification. Bacteria, seen only as causes of disease, were then and are now nearly always branded as "enemy agents." Note how they are "waiting to be conquered" by the "weapons" of modern medicine. It is ridiculous, of course, to describe them primarily in military, adversarial terms: most bacteria are no more harmful than air, nor can they, like air, ever be removed from our bodies and our environment. But many still erroneously believe that any bacteria, if present, should be eradicated. Bacteria now and even more in Wallin's day must be vanquished. How could they "inhabit" healthy tissue? Wallin's colleagues confused the map with the territory.

For most people today, life is readily divided into three categories: plants (for food and decoration), animals (like our pets, seafood, and us), and germs (to be vanquished). I can't remember when I realized that this idea is as dangerous as it is prevalent; certainly it was long ago. I try to replace this oversimplified cultural nonsense with concepts much closer to hard-learned scientific truths. Neither plant nor animal appeared on Earth until bacteria had undergone at least 2000 million years of chemical and social evolution. Indeed not only animals and plants but even fungi are new to Earth. Neither animal nor plant is an eternal category of classification. Neither was established once and for all by a divine mind with a Platonic bent. In addition to all plants and animals alive today there are at least three other forms of life. And here, in nonplants and nonanimals, lies the real biodiversity.

Animals and plants are far more similar to each other than they are to all the other kinds of Earth life! Thanks to electron microscopy and new molecular biological ways of studying details in all organisms, we understand the motley assortment of Earth life better than ever. Long-chain molecules such as DNA, ribonucleic acid (RNA), and protein permit us to study all life with a single standard of measure. The great animal-plant divide that has held sway since before Aristotle is crumbling. Radical revision overtakes our classification systems. Biologists explore the astounding details of the microbes, including their toughness in the face of adversity and their tendency to survive by making symbiotic evolutionary commitments.

My work on microbial symbionts prodded me to criticize and then to revise biological classification. For over two decades, Karlene V. Schwartz and I have collected taxonomic information from our colleagues and their scientific publications to reweave contradictory and limited biological classification systems into one consistent scheme. Our goal is a description that is as accurate and as useful as possible and that reflects evolutionary history. Our modern revision incorporates a two-tiered five-kingdom classification. The largest distinction in all life is that which separates the prokarya of the first tier—all bacteria composed of "prokaryotic" cells that did not evolve by symbiogenesis—from the second tier, which includes all the eukaryotic others. The eukaryotes, the organisms with nucleated cells, all evolved by symbiogenesis. This group includes protoctists, fungi, plants, and animals. This scheme, depicted in Figure 4, is increasingly useful.

Ernst Haeckel, an eclectic German scientist, respecting the protists, added their kingdom to that of plants and animals. But Antony van Leeuwenhoek (1632–1723), born more than two hundred years before Haeckel, discovered the microbial world. A cloth merchant from Delft in the Netherlands, van Leeuwenhoek spent his days, as I do, perusing the microcosm; he, however, built his own microscopes. In colloquial seventeenth-century Dutch he described rampant microbial life from puddles, pond water, the saliva of young beauties, and the diarrhea of drunks. Eventually his descriptions were published in London as letters to the Royal Society. Much later the Darwinian concept of evolution led European intellectuals to search for the common ancestors of life. Then van Leeuwenhoek's tiny beings became more than "freaks" to be shown off by the amateur naturalist focusing an eighteenth-century modification of the Dutchman's microscope. Microbes were seen by microscopists more and more as ancestral forms of larger life.

Microbes, of course, were not added to formal taxonomic categories until well after they were discovered. After Louis Pasteur's discovery of nefarious disease bacteria, the smallest organisms were named and forced into a classification scheme.

Long before that, in the third century B.C. Aristotle classified over five hundred animal species. Using only his keen eye, he of course saw no microbes, and he considered life's categories to be fixed and unchanging. Still, some of Aristotle's classifications correspond to our modern ones. Dolphins, for example, he grouped with land mammals rather than with fish. Later, the Roman scholar Pliny (23–79 A.D.) in a thirty-seven-volume *Natural History*, attempted to catalog all living beings ever reported. Relying on many sources of information, Pliny included unicorns, flying horses, and mermaids in his survey. During the Middle Ages and into the Renaissance, as descriptions of beings were reported in travel narratives, "bestiaries" were written and profusely illustrated. Elephant skeletons were taken as evidence of monstrous humanoids, and the fossil teeth of shark were interpreted as remnants of slain dragons.

Taxonomy became more reliable when in 1686 an Englishman, John Ray (1627–1705), published a compendium of thousands of different species of plants. In 1693 he issued his classification of animals, arranging them according to differences and similarities in their bodies: hooves, claws, teeth, and other features. Reflecting a growing distrust of the rumors, fables, and fantasies reported as fact in bestiaries, Ray claimed that fossils were remains of plants and animals that no longer existed.

The most complete preevolutionary classification scheme was published by the famous Swedish botanist from Uppsala Carolus von Linné (1707–1778). Writing his Latin name, Linnaeus, he invented what came to be called binomial nomenclature. He gave every life-form two names, both of which were usually derived from Latin or Greek. The "first" name signified the group to which the organism belonged, its genus, and the "last" name its species. These names were and still are written in italic type, with the genus name capitalized. The Linnaean system is today crucial to biological knowledge. Internationally all biologists recognize these two names as genus and species. Even Japanese and Chinese books, and Russian texts that use the Cyrillic alphabet, write species and genus names in Latin italics. No matter one's native language or region of origin, all authors and naturalists know the Linnaean names refer to the same species of organism. The genus is the higher, more inclusive taxon. The species is the smaller, less inclusive grouping.

All dogs, for example, belong to the genus *Canis*. The species of domesticated dogs is *familiaris*. Wolves are *Canis lupus* and coyotes *Canis latrans*. Humans are *Homo sapiens*. This classification Linnaeus applied only to our bodies; he considered our souls outside the properly classifiable scheme of nature. The only other members of the genus *Homo* today are extinct fossil humans like *Homo habilis*, *Homo erectus*, *Homo sapiens neandertalensis,* and the newly discovered *Homo sagittarius*.

Linnaeus also grouped genera into higher taxa called orders, and orders into classes. The French anatomist Georges Cuvier (1796–1832) later arranged the orders into "embranchments," large groupings corresponding to today's phyla. Cuvier, whose work was highly important to the collection at the National History Museum in Paris, extended the Linnaean classification to fossils. Both Cuvier and Linnaeus believed that all species were eternally separate forms created by an omnipotent God. Cuvier considered fossils to be evidence of past life that had disappeared with the biblical Flood and other catastrophes. He thus admitted that some animals had become extinct. But he saw no evidence of the creation of new forms of life since God's creation of the world. Although not evolutionists, Linnaeus and Cuvier, with their close attention to detailed relationships among the living, were fine scholars. Their works have lasting value. They enhanced what became the main lines of evolutionary thought that followed in the late nineteenth and twentieth centuries.

Ernst Haeckel (1834–1919), the brilliant German investigator of the natural world, was one of the first to champion

Darwin's evolution. He saw that the concept of evolution posed a problem for the ancient plant-or-animal dichotomy. Haeckel was not correct in all his assertions: he believed that life, even today, evolves from nonlife. Our ultimate ancestors—some, he claimed, still dwelling on the sea bottom—were strange beings from which both plants and animals originated. He asserted the ancestors were neither.

Haeckel extended, popularized, and systematically applied Darwin's ideas of evolution. He described many new beings. He illustrated beautiful ocean floaters, plankton. He was the first scientist formally to relocate tiny marine dwellers, the radiolaria and foraminifera, neither plant nor animal, in their own kingdom. Haeckel bravely bestowed upon them the appellation "Kingdom Monera," meaning primitive units. The borders of Haeckel's Monera kingdom fluctuated during his lifetime. He sometimes included amebas and slime molds, now classified with protoctists. In some works he even included sponges, now considered animals, in his Monera. Through the many editions of his many books, Haeckel was consistent in his rejection of the ancient two-kingdom classification system. He thought, as I do, that the rigid plant-animal dichotomy contradicted new knowledge and interfered with an understanding of the evolutionary history of life. Unlike Linnaeus, who classified ten thousand species of God's creation, Haeckel was a Darwinian evolutionist.

A biology teacher in Sacramento, California, Herbert F. Copeland (1902–1968), elaborated Haeckel's scheme. In 1956 he published a little known book dividing Haeckel's Monera into two kingdoms.[2] The category "Monera" he reserved for bacteria, noting the lack of nuclei in their cells. The second kingdom, the Protoctista, he took from the British naturalist John Hogg's work of 1860. Copeland placed all microbes whose cells possess nuclei in kingdom Protoctista. He included traditional protozoa, the water molds that repro-

duce by forming spermlike swimming cells, and algae of all
kinds in Hogg's protoctist kingdom. Many other strange
groups of nucleated microorganisms or slimy larger life-
forms he unilaterally decided to include as Protoctists. He
collaborated with no one and asked no one's permission to
change classifications to suit his scheme. The most significant decision he made was to include molds, yeast, mushrooms, and all other fungi as Phylum Inophyta in his version
of the protoctist kingdom.

Three years later, a Cornell University professor, Robert
H. Whittaker (1924–1980), carefully developed Copeland's
prescient, but virtually ignored, four-kingdom taxonomy.
Whittaker, who founded the field of community ecology in
North America, had studied the pine barrens of New Jersey
for many years.[3] He realized that fungi in the pine barrens,
mostly root-connecting mushrooms and never photosynthetic, were so unlike plants that they belonged in a separate
kingdom. Whittaker established five kingdoms: fungi, plants,
animals, Hogg's protoctists (presented as protists, the smaller
members, which don't belong to any other kingdom), and
Haeckel's Monera (bacteria). Whittaker, like Copeland,
noticed that the first four of these groups are eukaryotes:
their cells always contain nuclei. Only members of the last
kingdom lack nuclei and therefore are prokaryotes. Members
of this group, all bacteria, remained in Haeckel's Monera
after Copeland removed all nucleated organisms.

Like Copeland and Whittaker, Karlene Schwartz and I were
frustrated by our colleagues' inconsistences, contradictions,
and confusions. We found it difficult to teach stupidities of
botanical and zoological classification to our students. The
botanical was, and often still is, irreconcilable with the zoological. We needed to use and teach one consistent, comprehensi-
ble taxonomy that made sense. We have worked for years to
collect the work of many scholars and investigators: botanists,
zoologists, microbiologists, protoctistologists, mycologists,

algalogists (=phycologists), and others. We want one teachable scheme, an evolutionary system of classification, that reflects cell morphology, metabolism, genetics, and developmental biology. We saw the obvious: although plants and animals have different strategies of survival, they share great structural similarities. Both are made of cells containing chromosomes inside membrane-bounded nuclei. Both produce egg, sperm, and embryo. From the first time we read about it in *Science* magazine in 1969,[4] Whittaker's five-kingdom system seemed to us the best reflection of the evolutionary groupings of life's vast diversity. But all large life has direct ancestry among microbes: giant kelp came from tiny golden algae, slime molds evolved from amebas, and the ancestors of great green seaweeds were microscopic chlorophyte algae like many still alive today. The large organisms cannot be separated from their tiny close relatives. Hence, following Copeland, Karlene and I resuscitated John Hogg's inclusive term *Protoctista;* we use it in an expanded, evolutionarily robust version of Whittaker's "Protist" kingdom. We retain the informal term "protist" to refer to small members of the Protoctista (see Figure 5). All protoctists evolved, ultimately, from bacterial symbiosis. Some protists are unicellular; others have a small number of cells. Any free-living ameba, for example, is a protist. Amebas, ciliates, algae cells, along with seaweeds and the colonial form of amebas, slime molds, are all Protoctista. *Proto-* is from Greek for first, as in *protozoa.* But unlike "protozoa," the terms *protist* and *protoctist* (which from *proto-* and *ctista* means "first established beings") have no zoological connotation. I call the protoctists "water neithers." Some live in puddles, some in tree holes, others in lakes; still others float in the ocean. Although all are aquatic beings, none is either animal or plant. Although animals evolved from some of the protoctists (the zoomastigotes), plants from others (the chlorophytes), and fungi from still others (the chytrids), no protoctist is itself an animal, plant, or fungus.[6]

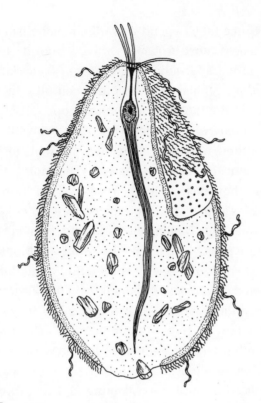

FIGURE 5

Mixotricha paradoxa, a protoctist: example of an individual composed of at least five kinds of organisms

We are persuaded that viruses do not belong in any of the five kingdoms. They are not alive since outside living cells they do nothing, ever. Viruses require the metabolism of the live cell because they lack the requisites to generate their own. Metabolism, the incessant chemistry of self-maintenance, is an essential feature of life. Viruses lack this. Through ceaseless metabolism, through chemical and energy flow, life continuously produces, repairs, and perpetuates itself. Only cells, and organisms composed of cells, metabolize. Whether capable of invading plants, animals, fungi, or protoctists, any virus outside the membrane of a live cell is inert. Viruses nevertheless are important to the story of life on Earth. Since they depend upon the metabolism of others, the first viruses most probably evolved from bacteria. They probably started as irradiated portions of bacterial cells living in sunlight. Some viruses, quite complex in structure, may look, under an electron microscope, like miniature robots or hypodermic needles. How they evolved such complex forms though lacking their own metabolism is a question I do not pretend to answer. The point that bears mentioning, however, is that viruses are no more "germs" and "enemies" than are bacteria or human cells. Viruses today spread genes among bacteria and human and other cells, as they always have. Like bacterial symbionts, viruses are sources of evolutionary variation. Populations of the virus-infected organisms are honed by natural selection.

Viruses, like all forms of cell-based life, produce problems when they overgrow their habitats. The high-profile Ebola and other viruses are blamed for the wreaking of havoc in certain populations. Overgrowth of resources, viral or other, tends to be due to weakening and disruption of the ecosystem. We can no more be cured of our viruses than we can be relieved of our brains' frontal lobes: we are our viruses.

The two-part (prokarya versus eukarya) five-kingdom taxonomy closely reflects evolutionary history. Thus it is far superior to the misleading old plant-animal dichotomy. Bacteria evolved first. They branched into many different forms: red, purple, and green; fermenting, photosynthetic, and respiring; sulfide-producing and oxygen-producing; oval, eel-shaped, and rod-shaped. Even large treelike bacteria evolved. But bacteria did not only diversify. They also invaded and came to dwell inside each other. They swarmed around sources of food, including other bacteria. Having neither immune systems nor rigid external barriers, in attempting to feed they merged internally, and with and without their viruses, they exchanged genes. Survivors of thwarted aggression formed uneasy truces. Merged, formerly independent bacteria became new kinds of complex cells. These complexes became protists as speciation began. Tiny protists and their colonies produced a huge and diverse group of organisms. Some 250,000 species of protoctists are alive today. Even more are extinct. Their tiny remains, microfossils, inform us of their former existence.

The protoctists are nucleated microorganisms that underwent many evolutionary gyrations, including the alternation of mating unicells and adult-forming generations. The descendants of some protoctists eventually became sexually reproducing plants and animals. Symbiotic bacterial coevolution in our ancestors led to our protoctist antecedents. Each one of us is a massive colony of microorganisms. Former protists are now eloquently orchestrated animals with fancy tissues and organs. Karlene and I feel that our modified "Whittaker five-kingdom taxonomy" faithfully reflects the evolution of protoctists from symbiotic bacteria, and of animals, plants, and fungi from protoctists.[5]

Carl R. Woese, of the University of Illinois at Urbana, along with his colleagues, has proposed a radically different, three-part taxonomy, consisting of Archaea (at first called archaebacteria), Eubacteria (all other bacteria), and Eukarya (all nucleated forms of life). He calls these groupings "domains." His Eukarya classification, all nucleated organisms, is the same as ours. Yet he affords Eukarya the same status as each of his two groups of bacteria. The four distinct eukaryotic kingdoms (protoctists, fungi, animals, and plants) are, for Woese, clumped together in one. In my view, Woese obscures rather than illuminates the critical distinction between prokaryotes and eukaryotes, between symbiogenetic and nonsymbiogenetic life.

Woese's two other domains, as their names suggest, are bacterial. Comparing RNA, one of the crucial long molecules present in all life-forms, Woese uses differences in the sequence, the order of RNA's chemical bases, to classify all organisms. After collecting data from nearly a thousand kinds of life, Woese classifies the different prokaryotes as either Archaea or Eubacteria. Members of Woese's Archaea domain include some halobacteria, prokaryotes that require salt water, and most acid-loving, hot-sulfur-spring bacteria. Also in Archaea are placed all bacteria that produce methane gas. The name *Archaea* from Greek means "old" and implies these were the first to appear on Earth. All other bacteria Woese classifies as Eubacteria ("true bacteria"). The Archaea-Eubacteria distinction, determined mainly by gene sequence analysis, requires extraordinary technology to verify. Despite their differences, both Woese's three-domain system and the five-kingdom system are consistent with a symbiotic view of evolution. Both are far superior to the plant-versus-animal anachronism.

In Woese's three-domain taxonomy, molecular differences between the two kinds of bacteria are given more importance than the differences between a mushroom and a moose. To me this is ludicrous. The best evolutionist in the world today, Ernst Mayr of Harvard University, fully agrees with us. This Woese system, he recently wrote to me, is getting into all the biology textbooks. "I think I'll do a little screaming." He has published a technical paper about the similarities of archaebacteria to other bacteria and the differences between all the bacteria and the rest of life, nucleated life.

Bacteriophile though I am, I believe our newly modified two-tiered (Prokarya, Eukarya) five-kingdom scheme, published in 1998, is far superior to the three-domain classification. The main distinguishing characteristic of all life-forms, nonsymbiogenetic cells (Prokarya) versus symbiogenetic cells (Eukarya), is given highest status. Attention is next paid to how an organism develops: from spores (fungi), from an embryo surrounded by its mother's tissue (plants), from a blastula embryo (animals), or from none of the above (protoctists).

Like Mayr, I see many problems with Woese's less biological system. Woese's main criterion to distinguish all life is a
single gene, a certain piece of DNA that codes for one of an
organism's RNA molecules. The one RNA this DNA codes for
is a part of the small subunit of a cell particle, the ribosome.
Woese uses only this single gene, although even small bacteria have about five thousand genes in each of their cells. I see
this as misleading. Organisms must be classified on the basis
of their entire biology. Second, at least one protoctist (*Plas-
modium*, the malarial parasite) changes the sequence of this
RNA gene at different times during its life history. If the RNA
of a microorganism can change in just a few hours, RNA
sequence is probably not the best way of defining the greatest of all groups. The most inclusive taxon to which an
organism belongs needs to be based on far more. Most people
can easily distinguish among the four great kingdoms of
large organisms on sight, but few of us have access to
Woese's gene-sequencing methods. Furthermore, a system in
which fungi, plants, protoctists, and animals are merged into
one group obscures hard-earned knowledge. A taxonomic
scheme must be an information retrieval system.

I have other objections to Woese's three domains.
Prokaryotes, unlike animals, plants, and other eukaryotes,
frequently pass on their genes one at a time. Archaebacteria
and eubacteria trade genes with each other. These tiny organisms are very similar and belong together in one empire,
kingdom, domain, or whatever you want to call the highest
taxon. The body form, behavior, and development of any
life-form must be considered along with its molecules and
internal chemicals. Even if there were identical gene
sequences in the bark of a banana tree and the skin of a dog,
we would still classify a dog not with a banana but with
wolves and jackals. The bountiful tradition of taxonomy that
leads botanists to classify plants on the basis of flower parts
is augmented, not overturned, by new molecular biological

insights. Although I applaud Woese's enormous contribution to universal classification, I think his three-domain scheme goes too far.

The move toward evolution-based classification systems has taken hold. Bacteria evolved first. They diversified by branching. Then, through bacterial symbiogenesis, branches fused as protoctists emerged. From a rich ancestral stock some protoctists evolved into fungi, others into animals or plants. Ancient groups remain and diversify. New forms may prove to be transient or stable. Whereas all species tend to become extinct, the large groupings, whether called domain, kingdom, or something else, persist.

Any taxonomic scheme has problems. We tend to label and dismiss anything once we assign it a category. Our classifications blind us to the wildness of natural organization by supplying conceptual boxes to fit our preconceived ideas. They should reflect our study of nature. The two-tiered five-kingdom system will always need revision. Whatever its difficulties, it does not perpetuate the age-old errors of the "animal versus vegetable" dichotomy. We can group life into three or five or a million categories, but life itself will elude us.

....

CHAPTER 5

LIFE FROM SCUM

What mystery pervades a well!
That water lives so far—. . .
Like looking every time you please
In an abyss's face! (1400)

Whether bacterial or nucleated, the units of life are cells. All visible organisms are composed of nucleated cells, and the first nucleated cell evolved by bacterial cell mergers, as we have seen. But how did that elusive unit, the parent of all Earth life, originate? What accounts for the beginning of the ur-cell? How did the very first bacterial cell originate? This question is exactly equivalent to the question "How did life originate?" To appreciate SET, which only recombines, merges, and integrates exceedingly diverse bacteria, we first have to understand where these diverse bacteria came from. In short, we need to try to understand life from scum.

In search of the ecological setting of the earliest cells on Earth, every few years my students and I make a pilgrimage to San Quintín Bay, in Baja California Norte, Mexico. We seek the shifting shores of Laguna Figueroa, a lagooned complex festooned with salt flats. Here we find laminated, brightly striped sediments underlain by gelatinous mud. These colorful seaside expanses, called "microbial mats," enchant me—a living landscape just where the sea meets and rolls back and forth over the land. Luckily for our studies, the scene is inhospitable to the vast majority of large life-forms, humans included. I put my hands in the mud of fragrant microbial tissues and whiff the exchanging gases. Here, as in the human sphere, but neither by commandment nor of necessity, death is part of life. Population growth potential is alternately checked and realized. These seaside communities have persisted for over three billion years. Many inhabitants die daily but the community itself never overgrows its bounds. This is an evolutionary Eden more primal than the greenest grassland. Here, in this Earth tissue, animals and plants are all but absent. Even protists and fungi are rare. Mostly bacteria thrive. Standing at the microbial mat, I feel privileged. I delight in escape, thrilled to abandon the urban sprawl of human hyperactivity and exhilarated with the freedom to contemplate life's most remote origins.

The origin of life is a mythical concept, not in the sense of being untrue but rather in stirring a deep sense of mystery. Even scientists need to narrate, to integrate their observations into origin stories. How did the earliest life, the first bacterial cell, begin? How did the ur-bacteria differ from the environment from which they are alleged to have emerged? Not only is such a question within the province of scientific inquiry, but an adequate answer is essential for SET. We must know how bacteria started and what they became before we can understand how separate bacteria merged to create our cells. Answers to the "origin of life" problem are woven from the lifework of scholars in many nations. The scientific story of the first life on Earth is the least parochial of the world's origin myths. It is freely available to all who care to learn about it.

The properties of minimal bacterial life, first life, can be inferred by several approaches. First, one compares all living beings to see what they share in common. Common and absolutely required aspects of all life are deduced to have persisted from our earliest bacterial ancestor. The chain of life has not once been broken since its formation.

A second approach to origins is via paleobiology: studies of microfossils, the remains of early life. Ages can be assigned to some microfossils. They may be dated by direct measurement of the age of the surrounding volcanic rocks with which they are associated.

A third avenue to illuminate life's origins attempts to remake a cell. The minimal ancestral state of nature is chemically imitated in the laboratory. Some components of life have been synthesized from simpler compounds in this way, but so far nothing close to a laboratory re-creation of a bacterial cell has been achieved. Of course, even if it were, we could not conclude that our clumsy imitation was the way cells really originated in the first place.

Through a combination of methods, I have come to agree with other scientists on the most likely, and investigable, scenario for the origin of the first cell on Earth. Before cells there were cell-like systems. Today, no piece of DNA, no gene, replicates outside the cell of which it is a part. Nor does any virus make more of itself without inhabiting a live cell. The bacterial cell, today's minimal unit of life, self-maintenance, and reproduction, is where we must begin.

No one claims to have "solved" the origin of life problem. Yet although we cannot create cells from chemicals, cell-like membranous enclosures form as naturally as bubbles when oil is shaken with water. In the earliest days of the still lifeless Earth, such bubble enclosures separated inside from outside. As Harold J. Morowitz, distinguished professor at George Mason University, Fairfax, Virginia, and director of the Krasnow Institute for the Study of Evolution of Consciousness, argues in his amusing mayonnaise book,[1] we think that prelife, with a suitable source of energy inside a greasy membrane, grew chemically complex. These lipidic bags grew and developed self-maintenance. They, through exchange of parts, maintained their structure in a more or less increasingly faithful way. Energy, of course, was required. Probably solar energy at first moved through the droplets; controlled energy flow led to the selfhood that became cell life.[2] By definition, the most stable of these droplets survived longest and eventually, at random, retained their form by incessant interchange of parts with the environment. After a great deal of metabolic evolution, which I believe occurred inside the self-maintaining greasy membrane, some, those containing phosphate and nucleosides with phosphate attached to them, acquired the ability to replicate more or less accurately.

However the first bacteria happened we can only guess. Yet the oldest fossils we have today are interpreted to be remains of fossil bacteria. Over 3.5 billion years old, probably the best preserved come from southern Africa. The very existence of these Swaziland microspheres, as they are called, shows us that life was already thriving, reproducing, and growing, just 1.1 billion years after the Earth's origin as a solid rocky body with atmosphere and ocean. No one today doubts that life on this planet is very old. Since the universe itself, exploding into existence from the "singularity" of the Big Bang, is usually dated at only 12,000 to 15,000 million years, the greater-than-1000 million-year tenure of life on Earth suggests life's presence for a quarter of the time span of the universe. Nor does anyone doubt that the earliest life was neither plant nor animal.

As the material stuff from which all living bodies are made, we have in another sense been around since the origin of the universe. The matter in the bodies of all lifeforms, including, of course, mammals like us, can be traced to the carbon, nitrogen, oxygen, and other elements that were made in the supernova explosions of stars.

At first it may seem improbable to you that all life on Earth today, inhabitants of cities, jungles, oceans, forests, and grasslands, is the progeny of an ancient bacterium. How could one or a few bacteria have been so prodigious? But note that you, yourself, were once a single cell: the fertilized egg, the zygote, that reproduced by division to become an embryo in your mother's womb. Later you became a crying infant in her arms. If in nine months a single fertile egg can become a human, albeit a pudgy, defenseless, and uncoordinated one, is it not easily conceivable that all life-forms today arose from a single bacterium over 3,000 million years ago?

The smartest cells, those of the tiniest bacteria, about one ten-millionth of a meter in diameter, continuously metabolize. This simply means they continuously undergo hundreds of chemical transformations. They are fully alive. Recent work has revealed that the tiniest, most simple bacteria are very much like us. They continuously metabolize, using the same components as we do: proteins, fats, vitamins, nucleic acids, sugars, and other carbohydrates. It is true that even the simplest bacterium is extremely complex. Yet its inner workings are still like those of larger life. All the DNA of one of the simplest cells, a bacterium called *Mycoplasma geniticulum*, has already been sequenced. This means we know all the details of its genes. The more closely we study gene sequences and metabolism, the more we realize that all life since its origin has been similar to its brethren, all other life. *Mycoplasma,* like all the other bacteria, never stop using energy to take up food; balance their salts; make DNA, RNA, and proteins; and convert one chemical to another to keep themselves going. They emphatically differentiate themselves from their surroundings. The simplest bacterial cell on the early Earth, like the smallest ones today, already had integrity. The tiniest first bacterium was already so complicated that Sir Francis Crick, one of the discoverers of the structure of DNA, made an astounding claim. Crick wrote an influential book, *Life Itself,* arguing that life, because of its overwhelming complexity, must have been brought to Earth from outer space.[3] He suggested that bacterial life was sent here by an extraterrestrial civilization bent on seeding planet Earth. Crick, with uninterpretable seriousness, claims that just as a human gardener introduces seeds into the soil of her backyard, propagules were planted on Earth eons ago. This idea, called directed panspermia or "pangenesis" and put forth for centuries, that life came in seed form from outer space, seems to me to stem from ignorance of evolution on Earth.

To transfer the problem of life's origin to outer space is intellectually unsatisfactory. Why should it have been easier for life to originate elsewhere than on Earth? Wherever cellular life began, it faced the same problems of origin. The idea of spontaneous generation shows not the origin of species but the origin of the specious.

For generations, Europeans believed that life appears spontaneously in scum and muck. Rotting meat was thought to generate maggots. Old rags grew mice. Close observation and experiment, however, revealed intermediates. Maggots, we know, do not appear from messes no matter how chemically complex. They grow from sperm-fertilized eggs laid by flies. Yet maggots wriggling in stinking meat, in the minds of Louis Pasteur's predecessors, meant that life arose from the decay itself. In the 1860s Pasteur exposed boiled meat extract to the air. He used a long flask, whose thin, downward-pointing neck admitted air but not bacteria or any other propagule. Another flask, open to the air, putrefied with growing bacteria and fungi within a few days. Pasteur's "control," the closed flask with its spout facing down, has never spoiled. The uncontaminated broth is still on display at the Pasteur Institute in Paris. The last of believers in spontaneous generation, Pasteur proved dramatically, were incorrect. Puppies come from dogs and bitches; babies come from men and women. Flies come from maggots; mice come from inseminated mother mice. Like them all, microbes come from preexisting microbes, or at least one unisex parent microbe.

However, there is some irony in this tale: Pasteur, a serious Catholic, interpreted his finding, as we all still do, to mean that all life must arise from preexisting life of the same type. But for Pasteur this showed that evolution does not occur, and that only God made the many types of life. Today's scientists turn the argument around: all life came not from the hand of almighty God, they argue, but ultimately from first life, and first life originated from nonliving solar system matter. This is the irony.

Bacteria, Pasteur convinced us, are just as alive as we are. Bacterial presence is correlated with disease and food contamination. Pasteur's brilliant experiments enjoy a great legacy. He established the prevalent view: infectious, indeed near-diabolical, bacteria are "germs" that need to be destroyed. The great successes of modern medicine reinforce the idea of microbes as enemy. Cleanliness, sterilization of surgical instruments, and especially antibiotics are all described as weapons of war against microbial aggressors.

The more balanced view of microbe as colleague and ancestor remains almost unexpressed. Our culture ignores the hard-won fact that these disease "agents," these "germs," also germinated all life. Our ancestors, the germs, were bacteria.

Whence came the first bacteria?

Spontaneous generation, as Pasteur and others triumphantly demonstrated, does not occur nowadays. But this Pasteurian observation is misinterpreted by creationists and other dogmatists to argue that life *never* came from nonlife. Some information theorists suggest that the chance that life would organize itself from nonlife by random interactions of molecules is so improbable as to constitute a "mathematical proof" that the origin of life was divine. But to me their assumption that life originated as a result of molecules mixed at random is flawed.

Direct experiments on the "origins of life problem" began in 1953. Stanley L. Miller, then a twenty-two-year-old graduate student of the Nobel laureate Harold C. Urey at the University of Chicago, filled laboratory glassware with gases floating over a surface of sterilized water. For a week he exposed this miniature diorama of the chemistry of early Earth to periodic electricity, mimicking lightning. Through the technique of paper chromatography he separated some of the many organic compounds that spontaneously formed. He recognized among these alanine and glycine, two amino acids found in all proteins and in all cells of living bodies. Miller and Urey gleefully concluded that such "spontaneous generation" of the chemical components of life was a natural result of chemical interaction.

In space, or on the early Earth, we think organic compounds like those found by Stanley Miller spontaneously formed from simpler precursors. Miller and Urey, of course, could only guess about the chemical characteristics of the early Earth's surface environment. The gases Miller used in his glass apparatus—hydrogen, water vapor, ammonia, and methane—seem reasonable. These gases are all hydrogen-rich. Hydrogen, the main element in the sun, forms more than 90 percent of the matter in the entire universe. Miller reasoned that hydrogen probably was abundant on the inner planets early in the history of the solar system. Thus arose the notion of a "primordial soup," that life emerged from a "gemish," a complex structure like that which floated in or stuck to the sides of Miller's flask. Earth, we think, was steeped in organic compounds synthesized by sunlight and other energy sources long before life. A Miller-like experiment probably occurred on a planetary scale. If twenty-two-year-old humans like Stanley could produce amino acids in the laboratory in only a few days, why could not the laboratory of Earth, in an experiment over the course of a thousand or a million years, produce life?

More recent experiments confirm that precursors of life can be produced naturally in the laboratory under conditions simulating the environment of the early Earth. Molecules, however, do not combine at random: carbon, hydrogen, nitrogen, phosphorus, oxygen, sulfur, and the other elements of life interact according to the rules of chemistry. The science of heat and energy, called thermodynamics, has laws that molecules obey. Certain chemical reactions are far more likely than others; the notion that all chemical combinations are equally likely may be convenient for calculating life's improbability, but it is not accurate. Moreover, if we assume that life evolved from the beginning in cell-like lipid droplets, the odds for the emergence of startlingly self-maintaining systems increase. A tendency toward complexity ensues.

To me the most exciting research now under way concerning life's origins is that of my friend Harold Morowitz. To space, time, and causality in biology, Morowitz adds "memory." Biology, he claims, is the bridge between physics and history. The oldest rocks on Earth, including those from the Isua formation in Greenland, are nearly four billion years old. All life has chemical memory that cannot be dated by direct measurement. The metabolic memory of modern cells most likely even predates the most ancient rocks. Some metabolic pathways, such as the enzymatic steps leading from fatty compounds to steroids like cholesterol, Morowitz notes, are limited to animals. Others, however, are components of "primary metabolism": metabolic pathways common to all living beings. Because certain carbon-chemical metabolic pathways are absolutely necessary for all metabolism, the earliest ones that underlie the cellular phenomenon of self-maintenance and were present from the beginning, the chemical interactions of carbon, nitrogen, sulfur, and phosphorus upon which all metabolism is based must be retained in all cells at all times.[4] Any cell dies if its universally required metabolism is hampered by environmental restraints, lethal DNA mutation, or other interference.

Chemical systems in nature that become more and more complex, even if they are capable of making more of themselves, are not necessarily alive. Such systems are called *autocatalytic.* An autocatalytic system is a cyclical series of interlocking reactions whose end product is the same as its starting point. Some such reactions have been referred to as "chemical clocks" because they do not quickly reach a steady state; rather, they persist and repeat. The Belousov-Zhabotinsky system is a colorful series of self-sustaining reactions. Malonic acid is oxidized by bromate in a sulfuric acid solution containing cerium, iron, or manganese atoms. As these chemicals react with each other, and then react again and again, concentric and rotating spiral waves are produced that often endure for hours before reaching a final stable pattern. A thermodynamic analysis of these reactions interprets them as dissipative structures, as described by the Belgian Nobel laureate Ilya Prigogine. A *dissipative structure* is any system that maintains its function through assimilating useful energy and dissipating useless energy, usually heat. Reactions of dissipative structures share certain traits with life and the chemical systems that evolved into life. But all chemical systems, dissipatively structured or not, only continue to operate and make more ordered matter for a short time. Then they fall apart.

From thermodynamic analysis and scientific experience, we infer that perpetual motion machines cannot exist: although energy itself does not disappear, it is irretrievably transformed. Dissipated heat cannot be recovered. High-quality energy, energy that can do work, tends to disappear over time. Snowmen melt and do not re-form. Cups and glasses are broken far more frequently than they are put back together. Messing up a room is far easier than tidying it. In thermodynamics, messiness rules. That energy is lost and things fall apart, never to reunite, is an inescapable fact, a law, of nature. Life, with its complex order, does not violate the thermodynamic law of inexorable tendency toward disorder. Life always requires its specific source of high-quality energy. Sunlight moves through life, empowering cyclic work, in much the same way that chemical energy channels through a Belousov-Zhabotinsky reaction. But because cells grow and reproduce to form more cells like them, once life evolved the life chemistry never ceased. Cyclic life, if provided a continuous source of energy and nutrients, will indefinitely make more of itself. Chemical systems lack selves: they can't make more *selves*. Life has always been identified as a series of selves—organisms or cells. These must expend energy to continue to exist but they do it in unseverable connection to past life. Life has been, since inception and with no discontinuity, chemically connected to its past.

Morowitz points out that the cumulative metabolic chart of living organisms, worked out by hundreds of scientists, mainly since the beginning of this century, is one of the greatest and most underappreciated intellectual achievements of humankind. Several Nobel prizes have been awarded for deciphering significant fragments of metabolism, the intertwined chemical reactions of cells. Only Morowitz, as far as I know, tries to organize the massive details of metabolic information into a single coherent whole, a lens to peer into life's ancient history.

Because life is intrinsically a memory-storing system, some scenarios advanced to explain its origin seem unlikely to me. Crystals, glasses, coacervates, clay, and iron pyrite (fool's gold) have all been claimed to be keys to the earliest prelife chemical systems. Advocates tout rock crevices or clay particles as the sites of the origin of life. Cavities filled with fluid exist in the membrane-bounded cells of nearly all living beings. Similar cavities, chemical bags called *liposomes,* also arise naturally. Such liposomes, membrane vesicles, appear spontaneously in so-called origin-of-life experiments. These sorts of droplets appear to me to be far more likely to represent life's original natural architecture than iron pyrite, clay, or glass. A principle of life's continuity, of life's memory, can be invoked here. I think the proverbial primordial soup of free-floating DNA or RNA never existed, because nucleic acids (DNA, RNA) are far more easily destroyed than they are spontaneously formed. Membranous structures are the sine qua non of life. Today the membrane-bounded entities with identity and integrity are cells. Life arose in its cellular wholeness. The cells of today are, as Morowitz says, "virtual fossils."

In all of today's cells, genes are made of DNA. RNA, quite similar to DNA, is needed by all cells to synthesize protein. The precise amino acid sequence gives a protein much of its structure, and so determines what it will do, just as the sequence of letters gives the written word its meaning. Proteins exist in many sizes and shapes with hundreds of functions. Some proteins pump ions: sodium, hydrogen, phosphate, potassium, and others. Other proteins, attached to pigments, provide energy absorbers in dark eyes, spotted skin, green cyanobacteria, and algal plastids. Muscles are primarily protein; blood, skin, and tongues are complexes of many proteins packed in cells.

Cells work by a two-part system. First, they copy or "replicate" their genes. This gene-making step is DNA synthesis. DNA is copied, and one copy of the hereditary information is put on reserve. The other copy is "translated": identical base sequences of selected parts of the genome are made into RNA. Inside the cell, on tiny "factories" called ribosomes, the RNA directs the fabrication of long-chain proteins. Proteins of different kinds, some three thousand to ten thousand per cell, form most of an organism's body. Growth ultimately means protein synthesis (and, of course, water uptake). Together in a fluid-filled membranous bag DNA, RNA, and protein make the self-maintaining structure of cells. The RNA molecule, however, is more versatile than its DNA complement. Given the appropriate chemical milieu, but without any protein, RNA can autocatalytically make more of itself. DNA, on the other hand, requires both RNA and enzyme proteins to complete its work of replication; DNA by itself is dead. The capacity of RNA both to accelerate chemical reactions and to replicate suggests that RNA preceded DNA in the history of life. We can use RNA as an index of proximity to prelife. Nothing smaller than a live cell maintains its identity and produces greater numbers of itself. From the beginning, life was a cell, a mutual interaction between gene molecules (like RNA) and the oily membrane that segregated them from their environment.

The physicist Freeman Dyson suggests that the first life arose from a molecular symbiosis, a coming together of relatively amorphous "protein creatures" and the supermolecule RNA. Like most of us Dyson is impressed with RNA as a supermolecule, which, like DNA, replicates itself but which, unlike DNA, also directs amino acids into protein sequences. Although I think Dyson misuses the word *symbiosis,* his story of independent development of macromolecular sequences followed by strong interaction has merit.[5]

As Dyson knows, the unique talents of the RNA molecule are confirmed by laboratory experiment. In the late 1960s, at the Göttingen Institute in Germany, the Nobel prizewinning physicist Manfred Eigen showed RNA molecules that replicate in test tubes by themselves. He and his colleagues, including Don Mills at Columbia University and the late Sol Spiegelman of the University of Illinois, Urbana, showed that test tube RNA could mutate into new RNA molecules that replicated more rapidly than their "parents." Test tube

RNA molecules by themselves, like viruses, proteins or DNA in solution are dead. Yet molecular systems can proliferate and mutate in the test tube when provided proper support.[6]

Thomas Cech of the University of Colorado and Sidney Altman of Yale University, both very young in the early 1980s, made the key discovery. Certain RNA molecules not only replicate but act like proteins: they splice themselves. They thus rearrange their own molecular form. Cech and colleagues proved without a shadow of a doubt—that is, without contaminating proteins—that RNA behaves as the kind of protein that can rearrange and reorganize genetic material. This kind of RNA is dubbed a "ribozyme." Bits of RNA, with small spare parts (chemicals called ribonucleotides), given ribozymes evolve by themselves in a test tube. Let me emphasize that the RNA mix—whether or not enclosed in a liposome—is still not a cell. RNA and/or DNA molecules in a bottle are in no way alive. If left unmanipulated, neither test tube RNA nor DNA alone is even a virus. They are food for enterprising bacteria, protists, and fungi. But RNA molecules do evolve in the test tube, suggesting that biochemical evolution may have preceded life. Gerald Joyce and Jack W. Szostak, at the University of California, San Diego, engineer ribozymes that accelerate RNA replication. A Harvard University scientist, Wally Gilbert, a Nobel laureate, coined the catchy phrase "RNA world." Gilbert drew attention to RNA potency, suggesting among other good ideas that RNA acting as replicating ribozyme formed the kernel of the first living cell. I quite agree with Wally that RNA metabolism-accelerating reactions and replicating molecules preceded any DNA-based molecules. However, *both* RNA and DNA types of metabolism, as Morowitz emphasizes, live *inside* cells.

No life-form exists outside a self-maintaining, self-reproducing cell. The most stripped-down minimal form of life on Earth is still extraordinarily complex. Just a tiny membrane-bounded sphere, a wall-less bacterial cell requires a cadre of molecular interactions, more than fifteen kinds of DNA and RNA, nearly five hundred different types of protein and usually closer to five thousand kinds. RNA by itself, DNA by itself, any virus alone, is not alive. All living cells, even in principle, are much more complicated than any gene or virus. Cells interchange their parts; they maintain themselves continuously from nutrients and energy drawn from their surroundings. I agree with Morowitz: the first life-forms were membrane-bounded self-maintaining cells, like those still alive today.

Using the principle of continuity, Morowitz casts autotrophs, bacteria that make their own food and generate their own energy from inorganic materials, as the original membrane-bounded cells. Photoautotrophs do not have to eat; they use sunlight for energy. Chemotrophs do not have to eat; they use hydrogen-rich chemicals without the aid of light for energy. Both photoautotrophs and chemotrophs derive carbon from the atmosphere's carbon dioxide (CO_2). Neither eats organic compounds; that is, neither eats food. Plants; cyanobacteria; and ammonia-, sulfide-, and methane-oxidizing bacteria are all autotrophs. The opposite of an autotroph is a heterotroph: any organism (herbivore, algivore, bacterivore, carnivore, or cannibal) that eats food. "Eating food" is the same as taking in preformed organic matter. All heterotrophs eat organic molecules made by autotrophs. Autotrophs "eat" air for food. They "eat" sunlight or employ the mephitic power of hydrogen-rich compounds, such as hydrogen gas (H_2), methane (CH_3), hydrogen sulfide (H_2S), or ammonia (HN_3), to make more of themselves. The energy of autotrophs is the same as fire: the hydrogen-rich compounds react with oxygen. Morowitz thinks the autotrophs still close to Earth's original nonliving geochemistry were the original type from which the rest of us sprang. Reasoning that autotrophs are closer to life's original thermodynamic cycles,

Morowitz postulates that the chemoautotrophic way of life even antedates photoautotrophic.

In seminars, Morowitz unveils the chemical layers of the glass onion step by step. We just converse and listen to him. My students and I like it when we help Morowitz and others trace life's history from its living present to its inert chemical past. We need to know the transition from prebiotic chemistry to cell-based life so that we can figure out how organelles evolved. Did they emerge directly from prelife and complexify, or are they stripped-down bacteria? I think and read about this extensively, but what I do with students and colleagues, in the laboratory and the classroom, is different. We deal directly with life: microbes and other live cells and their parts. Bacteria, protoctists, plants, or fungi are our objects of study. With my students and colleagues I trace life's history from its microbial antecedents. We observe growth and reproduction; we spy on protoctist sexuality and physical maturation; we measure the responses of bacteria and protists to environmental "insults." We are especially concerned with these microbes' behavior, rich social lives, and interaction with sediments as they form persistent community structures.

The origin of cells from scummy chemicals may have occurred once or many times. In any case, the first cells in our lineage were membrane-bounded, RNA- and DNA-based, self-maintaining protein systems. In details of cell structure and metabolic behavior they very much resembled us. Their material constituents continuously exchanged with their external environment. They vented waste as they acquired food and energy. Their patterns persisted as they replenished their innards with chemicals taken from the surroundings. Indeed, metabolizing ancient bacteria were so effective at remaking themselves when threatened with disintegration and thermodynamic demise that the insides of our bodies today are chemically more akin to the external environment of the early Earth, in which life originated, than they are like our present oxygen-rich world. Life, always made of cells that grow and divide, literally has preserved its past as chemistry. The book of life is written in a language that is neither mathematics nor English: it is the language of carbon chemistry. "Speaking" the language of chemistry, the bacteria diversified and talked to each other on a global scale. Those that swam attached to those that degraded glucose, the sugar, and so generated power for swimming. The swimming, glucose-degrading partnership led to protists. The rest is history—my SET vision of history.

....

CHAPTER 6

When a Lover is a Beggar
Abject is his Knee —
When a Lover is an Owner
Different is he — (1314)

Sex is tricky. The essence of sex throughout life's evolution is the coming together of complementary gendered cells in fleeting or prolonged union. The core of sex, biologically, is gender attraction strong enough to lead to recombination of genes inside mated cells. New live beings emerge that differ in their genetic constitution from the gendered cells that met and recombined. Half their genes come from the egg, half from the sperm cell in animal-style sex.

Bacteria pass their genes with abandon as one bacterium donates its genes to another. No fifty-fifty contribution exists for bacteria. Bacteria literally pick up genes, usually a few at a time. The donor may mate when "he" physically contacts "her," a live bacterial recipient. "She" looks just like "him." Or gene uptake may be a casual pickup; the recipient may just grab genes shed earlier when some dead donor left them in the water. Picked-up genes may be for vitamin production, gas venting, or other traits that increase the chances of survival. Sometimes they code for proteins that permit the recipient to detoxify life-threatening poisons. Bacterial sex is always one-sided. Genes and only genes may pass into the recipient cell from anywhere: the water, a virus, or a donor dead or alive.

Human sex, like that of all animals, is a shockingly different process. This cell-fusion sex, called meiotic sex, is the same in animals and plants. Meiotic sex is not optional. Plants and animals do not "drink" genes from DNA-rich waters. Rather, sperm cells from the male parent are attracted to egg cells from the female and fusion ensues. In fertilization an act of embryo development is induced. A new single cell, the fertile egg, the zygote, whether animal or plant, now has two sets of chromosomes, whereas the original egg and sperm had only one. The fertilized egg then grows by division of its doubled cells. One divides into two, two into four, four into eight, and so on. The cells all stick together as together they form the embryo. The fertile egg becomes a recognizable member of its species. The new tadpole, moss stalk, seed plant, reptile, screaming infant, or other youthful multicellular being that develops from the embryo is living proof that animals and plants are more closely related to one another than they are to the rest of life. The sex lives of all the many beings that never form embryos differ profoundly from those of plants and animals. Bacteria and microbial descendants of their symbiogenesis (protoctists and fungi) have other sexual habits.

When bacteria wildly reproduce they need no sex to do it. The bacterial sex that responds to certain environmental contingencies is occasional. The sex lives of plants and animals, by contrast, are absolutely required for embryo making. Without sex the life history of animals and plants does not unfold. At the beginning of the life cycle of plants and animals the sperm nucleus permanently fuses with that of the egg. This fusion is reminiscent of cyclical symbiotic mergers: partners recognize each other.[1] They deploy cell emissaries.

Their cell membranes actually open up to passage of (at least) nuclei. The dissolved membranes re-form as the lover cells fuse. Open and vulnerable, the egg "knows" not to spill its contents and to let only the correct cytoplasm and nuclei enter!

Sexual processes, the merger of attracted beings, probably originated as did the early symbioses. In both sexual and symbiotic fusions, hunger was a likely primordial factor urging the desperate to merge. Cells that join in sex, however, by definition represent genes and cytoplasm from gendered individuals who are members of the same species.

In our book *The Origins of Sex*, Dorion Sagan and I argued that meiotic sex began long after bacterial sex as abortive cannibalism in certain protists.[2] To understand the convoluted history of sex, we declared, one needs to know protoctist biology.

We explained how both sexual and symbiotic mergers bring distant genes together within the recombined organism. Sex differs from symbiosis in that the cyclical fusion and later separation tend to be far more predictable, far less creative and casual than those in temporary symbiosis. In sex, offspring greatly resemble their parents, and gender differences are ritualized and predictable.

The bodies formed by symbiogenetic fusion, such as nodulated bean roots, green hydras, cud-chewing cows, luminous fish, and red algae, differ profoundly from each of the parent partners that fuse. Symbiogenesis is far more splendid than sex as a generator of evolutionary novelty. When the parents are extremely closely related to one another, for example, nonphotosynthetic red algae who live on (or rather, "off") their relatives, other photosynthetic red algae, sex and symbiosis are barely even operationally distinguishable.[1] But when the symbiotic parents of the mergers are distantly related—for example, bean plants and rhizobia bacteria or cows and their entodiniomorphid rumen ciliates—the products of these mergers are stunningly different from either parent.

Programmed death is a nonnegotiable consequence of the sexual mode of life. The great cycle in which males and females make sperm and ova with one set of chromosomes, only to have them come together again to make an offspring with two sets of chromosomes, is linked intimately with the imperative of individual plants and animals to die. All organisms, including bacteria and many protoctists, can of course be killed. Starvation, desiccation, and poisons are great killers. But death by destruction lacks a natural built-in timetable. Evolution of the protoctist ancestors to plant and animal bodies required sacrifice and loss; multicellularity and complexification ushered in the aging and death of individual bodies. Death, the literal dis-integration of the husk of the body, was the grim price exacted for meiotic sexuality. Complex development in protoctists and their animal and plant descendants led to the evolution of death as a kind of sexually transmitted disease. More than one billion years ago, when protoctists evolved by integration of bacterial symbionts into permanent and stable communities that became protoctist individuals, the kind of scheduled death that disturbs us today first appeared.

Many protoctists still exhibit variations on the strange theme of sex and death. I have watched *Stentor coeruleus* live and die. This blue ciliate, a large microorganism easily captured from unpolluted freshwater ponds and lakes, reproduces by single parentage. One individual grows and divides into two on a daily basis. In spring, in well-populated clumps, the blue stentors pair off. Microscopic orgies ensue as all stentors in the same bowl pair off and mate passionately. The pair members stick to each other for thirty-six consecutive hours! But both partners in the love match always die within less than a week. Sex, on the way out in this protist, is devolving.

Animals, all of them, an estimated 30 million species belonging to nearly forty phyla, revert to a single-cell protist-like stage in each generation. Protist-like sex cells enjoy sexual fusions when animal cells repeat the living styles of their ancestors. In animals, plants, and even fungi, sex is no dispensable option for staying in the evolutionary game. Living organization in these beings is generated by fusion, by the sex act. Mortality is the price they pay for fancy tissues and complex life histories.

The evolution of the great panorama of modern life on Earth from ancient symbiotic bacteria is great drama. Also wondrous, if less spectacular, is the development of mature adults from zygotes, fertilized egg cells. Evolution has continued for 3.5 billion years. The tenure of the human species on the "blue marble" in black space is fewer than 3.5 million years. A single human develops in fewer than 35 years. Time is concentrated, matter orchestrated, and a sentient being nearly invariably results. How does one pinpoint the beginning of a human life? The question is biologically absurd, totally artificial. The dating of the "start of human life" is simply convention. At public lectures I am often asked, "At what moment does human life begin?" Of course, it began, as all life did, at least 3.5 billion years ago! The question reflects a misconception. Once the genitals of the parents meet on time, conception occurs, to be followed by birth as the culmination of predictable in utero development. The development proceeds when an undulipodiated sperm meets a fat-filled ovum buffeted by the waving undulipodia lining the mother's fallopian tube. The haploid sperm who, unlike his fellow travelers, survives the journey combines with the haploid egg. The egg emits protein goo, welcoming the winner of the sperm race and barring subsequent callers. The winning sperm head, containing twenty-three chromosomes all carrying the father's genes, enters the large egg cell, whose nucleus bounds twenty-three very similar mother chromosomes. The two cells merge. This "fertilization" act makes a doubled cell, now having forty-six chromosomes. Of each pair of the twenty-three chromosomes one is maternal, derived from the mother. The other is from the father.

The mammalian fertilized egg resembles ancient protists, probably the first double-chromosomed beings to evolve. But unlike its ancestors, no mammal's egg lazily lingers in uni-cellularity. Nor does it use its undulipodium to swim through pond water to seek bacterial food. No, the zygote cell divides and the products of the divisions stay together. Human embryos result. A sphere of cells grows and differentiates to form tissues and eventually organs in a fishy-looking whole.

As cells continue to divide, grow, move over each other, and communicate, the embryo becomes a fetus. As organs develop, the embryo loses its fishy appearance, resorbs its gill slits and vestigial intrauterine tail, and begins to take the human shape. A fetus, open-mouthed, may even suck its in utero thumb. Unable to breathe air, it respires through its umbilical cord. Via the placenta, oxygen from the mother's blood is removed by the fetus. Delivered into the airy world, the infant, if female, is born with her lifetime allotment of unfertilized eggs already in place in her tiny ovaries. Each egg has twenty-three chromosomes, half the number of the rest of the infant female's body. Like the cells of both her parents, her body's chromosome numbers, at forty-six per cell, are already doubled. The chromosome numbers are already double in all the cells of boy infants' bodies. Males do not produce sperm with their twenty-three chromosomes until adolescence. Once adolescence strikes, the sperm-making cells in their testes, spurred by male hormones, continuously produce sperm. Sperm with only a single set of twenty-three chromosomes continue to form for up to ninety years. Without fertile union, sexual penetration, and offspring persis-tence, nothing counts on the playing field of evolution. If all our apish ancestors had not passionately mated to procreate us, we would have long ago become extinct. To be an animal is to be sexual.

How did multicellularity originate? Many of my colleagues and their readers take this as one of the big evolutionary questions. Animal bodies grow without sex by mitotic cell divisions of the fertilized egg. But how did the earliest animal bodies evolve? Nucleated cells, as protoctists, began cloning at least one billion years ago. Clones, organismic copies, resulted from cell division. Where did the sex requirement come from?

The oldest animals in the fossil record are found well preserved in more than twenty places worldwide: Sonora, Mexico; north of St. Petersburg on the White Sea of Russia; Ediacara in South Australia; Namibia in southwest Africa; scattered sites in China; and elsewhere. Here are preserved remains of animals between 650 and 541 million years old. Yet hundreds of millions of years before these animals lived, sexual protists already cloned and stuck together. "Individuals" with body parts differentiated into tissues were naturally selected. Different kinds of protoctist colonies, green algae, slime molds, nets, and many others, although amorphous early on, increased in complexity and individuality as they evolved. The individuals we recognize as plants, animals, and fungi today are best regarded as highly integrated protoctist clones. Naturally selected, they became new, larger beings.

Animals—coelenterates, brachiopods, and the thirty-five other phyla—evolved in the oceans about 500 million years ago.* Animals appeared long before the kingdoms of land-dwelling organisms, plants and fungi, made their appearance. Traditional geological timescales lump, and thus dismiss, all appearances prior to 541 million years ago as

---

* See note 5 in Chapter 4.

Precambrian. The Cambrian, named after the rocks in Cambria, an old name for Wales, in which skeletalized animals are first found, still captivates the paleontological imagination. How did animals evolve? One traditional argument, advanced by the geologist Preston Cloud (1910–1992), is that increasing atmospheric oxygen concentrations enabled animal life to evolve after thousands of millions of years of bacteria and algae. My close colleagues Mark McMenamin and his wife, Dianna, have shown that animal emergence is far from a simple one-cause phenomenon. Oxygen is necessary but not sufficient. Many environmental and genetic factors led to the first beings produced by sperm-fertilized eggs.

I applaud the McMenamins' analysis. Animal life was not restricted by anoxia, only to evolve suddenly when oxygen built up in the atmosphere. I do not believe oxygen ever limited any forms of life but the bacteria and a few obscure protists. Oxygen was necessary but not sufficient to ensure the expansion of animal life.

All animals are absolute aerobes. Their mitochrondria always demand oxygen, or they die. Oxygen in air, even in abundance, probably preceded the appearance of animals by more than 500 million years. Hard parts, such as the calcium phosphate skeletons of fish, the chitinous exoskeletons of arthropods, and the calcium carbonate shells of clams, snails, and other mollusks probably began as waste. Calcium ions, abundant in the seas, are poisonous inside cells. The concentration of calcium inside cells must be kept a thousand times lower than that in seawater, or the microtubules of mitosis are halted and growth ceases. Calcium extrusion, beginning as waste removal, evolved into an innovative style of recycling that led to systems of structural support. Teeth, armor shields, and skeletons evolved. Clever, economic, and feasible uses were modified as hardened calcium waste precipitated into phosphate-rich waters.

We humans may heed the lesson. We always create waste; life must excrete as it proliferates. No one is surprised when useful artifacts are fashioned from discarded automobiles and plastic party items. The metamorphosis of pollution has precedents. Prudently, unconsciously, we follow the lead of our remote ancestors.

By the mid-Cambrian period, life had magnificently propagated into animal forms that fascinate. *Hallucinogenia* is a monstrous little being so unlike modern life-forms that its discoverers were unsure whether its protrusions were protective spikes on its back or jutting legs. Another Cambrian era animal, *Pikaia*, was a soft-bodied darting swimmer, now thought to be the ultimate grandparent of all vertebrate animals.

Arguments among my friends don't cease. Preserved in sandstones dated from 650 to 541 million years old, well before the Cambrian period, large complex forms of life, still unidentified, even at the level of kingdom, abound at over two dozen sites worldwide. What are the Ediacarans? Are they a fossil fauna? Were they protoctists? Were some animals and some protoctists? What extinguished them? If they developed from the animal embryos (blastulas), we are without a trace of an embryo. All animals develop from the blastula. We have no way yet of knowing whether or not these strange beings ever did. (The chances of finding Ediacaran embryos have greatly increased since scanning electron microscopes have been applied to the study of phosphorus-rich sedimentary rocks from China called phosphalites.) Mark McMenamin persuades me that leaflike *Pteridinium* and three-armed *Tribrachidium* were not animals. He sees them as a distinct lineage in a gentler age, convergent to animals but not animals. They were protoctists that left no descendants (Figure 6).

Plants evolved from green aquatic protoctists, algae of the grass-green type. Familiar algae include iodine-rich rhodophytes (red algae), the edible wrappers for sushi rolls in Japanese restaurants. Because they never develop from any kind of embryo, seaweeds, such as rhodophytes and all

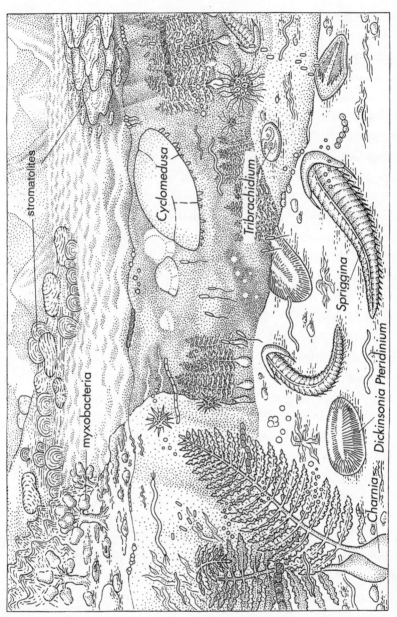

FIGURE 6

Ediacaran biota of the Late Proterozoic era

other algae, are not plants. Undifferentiated and resembling hypothetical ancestors to plants, certain algae provide an evocative series of evolutionary signposts. By their existence they show a plausible route from protoctist unicell to multicell descendant It happened over and over again to many forms of life. Unicells divided and the offspring failed to separate, as when large volvocine algae evolved from tiny *Chlamydomonas* ancestors. Pink snow algae of the high Rockies and the Alps and the red tinge on the snow at Hawley Bog near us in northwestern Massachusetts are members of this group of hardy life-forms. Although itself a loner, *Chlamydomonas* looks just like an individual cell of a rotating sphere, the far larger *Volvox*. *Volvox* consists of a layer of 500 to over 500,000 *Chlamydomonas*-like cells. Intermediate in this series is *Gonium*, an algal disk composed of four, eight, sixteen, or thirty-two green *Chlamydomonas*-like cells embedded in a translucent gel. Together such volvocine algae, *Chlamydomonas, Gonium, Pandorina, Eudorina,* and *Volvox*, represent variations on the theme of clones whose cells stick together and make multicellular individuals. One species, *Gonium sociale*, looks like a flat plate made of four parts, its cells. Any cell of *Gonium sociale* may break off, swim away, and start its own colony. A parent sphere of *Volvox* forms baby colonies inside itself. They release an enzyme that dissolves the gelatin holding the parent together. Out "hatch" several small offspring. When the season or lighting changes, *Volvox* changes its moods, becoming sexy. A sexy *Volvox* colony releases eggs. Another releases sperm or, if hermaphroditic, both egg and sperm will be shed from look-alike translucent green spheres. Undulating through the water, undulipodia forward, the green swimming sex cells superficially resemble single *Chlamydomonas* cells. But they are not individuals. They are sexual messengers, potentiators of multicellular individuals in the genus

Volvox. Probably the Ediacaran biota boasted similar life-styles of now-extinct multicells.

Working together, cells become colonies and colonies become individuals at ever higher levels of organization. Tissue differentiation requires repetition of past processes of development. In plants and animals fertilization by sex cell fusion represents the starting point of the life history. Even plants such as dandelions, strawberries, club mosses, and grasses, which are connected by underground runners and grow without benefit of sex, eventually turn sexy. Animals that have apparently given up sex, such as all-female groups of lizards, still undergo sex processes, meiosis and fertilization, at the cell level. A female cell with one set of chromosomes fuses with another female cell. The mother has cell sex with herself, to produce a fatherless zygote that develops into the animal embryo.

"We," a kind of baroque edifice, are rebuilt every two decades or so by fused and mutating symbiotic bacteria. Our bodies are built from protoctist sex cells that clone themselves by mitosis. Symbiotic interaction is the stuff of life on a crowded planet. Our symbiogenetic composite core is far older than the recent innovation we call the individual human. Our strong sense of difference from any other lifeform, our sense of species superiority, is a delusion of grandeur.

This delusion, I suspect, evolved from the need for "species recognition." We feel the need and the passion to breed and produce more people. That act of staying on the evolutionary playing field requires that we recognize potential mates of our own species. This sexual self-focus, however, obscures the larger symbiogenetic truth of our many-specied components. Multicomposition is our nature.

The zoological irony is that to understand the evolution of animal sex requires knowledge of the protoctists. But protoctists, often mistaken for miniature animals, are seldom studied by professional biologists. They suffer an indignity even greater than that of the bacteria demonized as disease germs: protoctists are ignored. Most people have heard of bacteria, even if only in the context of lethal sprays, poison powders, and antiseptic solutions. Among the estimated 250,000 species of live protoctists, only a few (amebas, slime molds, green algae, ciliates) even have names. Even these are mere curiosities for biology classes.

Animals and plants must sexually attract and fuse their nuclei to form embryos. Sex is not optional for them. But many protoctists never indulge in sex at all and are doing just fine, reproducing all the time.

Lemuel Roscoe Cleveland, while he was a professor of biology at Harvard University, published in *Science* magazine a very clear theory solving the problem of the origin of our kind of meiotic sex. As he studied live protoctists and saw their foibles, fumbles, and serious mistakes, he realized that fertilization began as an accident of desperation. Meiotic sex, as a strategy of survival, occurred in the aftermath of cannibalistic indigestion. Cleveland observed odd tensions in dying communities: one apparently starving mastigote devoured its neighbor; another squiggled out of the way of a hungry potential predator. Cleveland realized he was watching abortive cannibalism. Some cannibals ate and digested every last cell appendage of their victim brothers. Another might suffer indigestion and spare the nucleus and chromosomes of its intended meal. The two merged cells would form a new single cell with two nuclei and two sets of chromosomes. Cleveland, living daily in his microcosm, recognized the final cannibalistic truce. He noted that two such closely spaced nuclei fused. This was more than aborted cannibalism. Cleveland recognized it as the formal equivalent of fertilization.[3]

The complement to fertilization mergers is dissociation. Dissociation in sex requires meiosis. Meiotic sex, "animal"

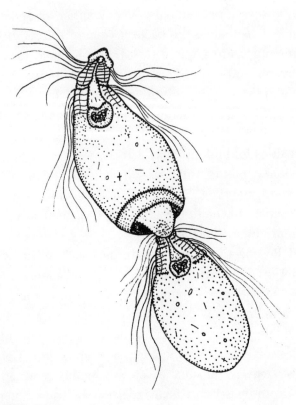

FIGURE 7

Mating protists: Abortive cannibalism in single-celled protoctists resulted in a truce called sex. Here two complementary-gendered Trichonympha mate. The female, with her fertilization ring, is penetrated through her posterior by the male.

sex—which also occurs in many protoctists, all plants, and most fungi—is the whole set of processes that halve the chromosome number per cell by special cell divisions. By meiosis, usually two divisions, the two sets of chromosomes per cell reduce to only one. These so-called haploid cells are then ready to detect, pair off, and fuse with other haploid cells to create diploid embryos.

Of course, nonphotosynthetic protoctists, whether or not they indulge in sex, always eat. Some will eat anything under stressed conditions. If desiccated, starved, irradiated, or otherwise in imminent danger of death, and attempting to save themselves by eating their fellows, they fuse rather than die alone.

Many freshwater protoctists seasonally mate before they die. The products of mating—zygospores, hystrichospheres, or other cells with doubled sets of chromosomes—tend to be tough-walled enough to survive winter or the dry season. The robust doubled cells at first do not grow; rather, they protect their genes and other cell parts through hot dry summers, freezing winters, or other hard times.

Survival by doubling began as an answer to environmental threats. How did this happen? If you were to eat your neighbor, undoubtedly a being genetically rather similar to you, but did not entirely digest him, in one comparatively rapid encounter you might nearly double your size. Like the bloated human, an engorged protoctist at the edge of hard times, with its two sets of chromosomes and twice the amount of cytoplasmic fluid, might be better able to withstand deprivation. Yet, in general, doubling the body, especially the chromosome numbers, is a hindrance. You'd probably yearn for relief from your double-monstrosity state. The original protoctist state is "haploid," which means organisms with only one set of chromosomes. In the haploid state a bewildering diversity of protoctists had long since evolved. The doubled set of chromosomes, advantageous during hard times, was complicated. The urge was to return to the haploid status quo.

Meiotic-fertilization sex probably first appeared a billion years ago. But, as Cleveland noted, meiotic, two-parent sex evolved only after the reduction-division of meiosis "relieved" diploidy. Eating-mating itself created irreversible gorging. As haploids ate each other they became diploids that ate each other, which became tetraploids, then octoploids, and so on. Chromosomes and bloat proliferated. The doubled cells with their extra chromosomes and other organelles were slowed down and even stopped in their everyday activities. Down's syndrome is one of many human genetic diseases caused by an extra chromosome or piece of a chromosome; it warns us of the dangers of chromosomal imbalance. Yet even today, many plants and nonmammalian animals tolerate extra chromosomes and chromosome sets. Breeders of Siberian irises, daylilies, and many other plants treat seeds with chemicals that inhibit microtubules of the mitotic spindle from forming and generate large cells with extra chromosomes. From such seeds beautiful and showy flowers may be generated. They inspire awe and profit; some continue to propagate in the desirably altered form. Before sex became ritualized, extra chromosomes and sets of chromosomes could be trivial, tolerated, debilitating, dangerous, or fatal. It all depended, as usual, on genetic and environmental contingencies. As Cleveland stressed, diploidy had to be reversed if doubled protoctists were to return to their streamlined haploid state.

Getting rid of cannibalistically acquired extra chromosomes, in Cleveland's analysis, was the first step toward meiosis. Meiosis, which produces sperm, ova, or haploid plant spores, reduces by half the number of chromosomes. The opposite of fertilization, meiosis turns diploids to haploids. The final refinement in the origin of meiotic sex was the perfection of the doubling/halving process so it occurred both on cue and without fail.

Because of Cleveland's prodigious insight, understanding the evolution of meiosis is easy for me. From my perspective the mitotic apparatus itself evolved from that ancient spirochete-archaebacterial fusion. The living vestiges of spirochete-archeabacteria long ago happily merged to act in concert. But the once-independent beings cannot be counted on to cooperate unfailingly. Their heritage reveals itself as they occasionally sneak a defiant act of reproduction on their own timetable.

Meiotic sex proper evolved in several lineages of protoctists. Ancestors became stuck in seasonal cycles of fusion and relief of mating and meiosis. When food, water, or other needs of living beings were scarce, cannibalistic "protomating" saved the day. Deprivation led to fusion and survival in the doubled mode. When environments were relatively lush, however, the older, streamlined, faster haploid cell organization was naturally selected.[4]

Sex, like symbiosis, is a matter of merging. But it is also a matter of periodic escape from the merger. Sex can be understood as a very special case of cyclical symbiosis: both sex (fertilized eggs, the zygotes) and symbiosis, merging of symbiotic partners, produce new beings. The act of mating (except in *Schistosoma*, trematode worms permanently stuck in the copulatory position as they abundantly make fertile eggs) tends to be brief. In animal and plant sexual mergings the new being is relatively long lasting, relative to the mating moment itself. In cyclical symbioses, such as luminous fish, nitrogen-fixing bacteria, and phosphate-moving fungi in plant roots, the merged beings also last longer than the dissociated partners. But cell symbiosis is a deeper, more permanent and unique level of fusion. In the great cell symbioses, those of evolutionary moment that led to organelles, the act of mating is, for all practical purposes, forever.

CHAPTER 7

....

ASHORE

Superfluous were the Sun
When Excellence be dead (999)

Until recently, when I visited the Star Trek commemorative
exhibit at the Smithsonian Air and Space Museum in Wash-
ington, I had never seen a single *Star Trek* episode. For ten
minutes, indolent curiosity, nostalgia for the 1970s, and the
crowds at my back induced me to watch it: very United
Statesian and very dated. I was struck by its silliness. The
lack of plants, the machinate landscape, and in the starship,
the absence of all nonhuman life-forms seemed bizarre.
Humans, if someday they trek in giant spaceships to other
planets, will not be alone. In space as on Earth, the elements
of life, carbon, oxygen, hydrogen, nitrogen, sulfur, and phosphorus and a few others, must recycle. This recycling is no
suburban luxury; it is a principle of life from which no technology can deliver us. Human voyages into deep space
require ecosystems composed of many nonhuman organisms
to recycle waste into food. Only very short stints in constant
contact with mother Earth are possible in the absence of
"ecosystem services."

An ecosystem is the smallest unit that recycles the biologically important elements. Carbon dioxide is "fixed," chemi-

cally converted to food and body (organic carbon). Organic carbon is respired, is reacted, is degraded or transformed to different kinds of organic matter. Eventually someone's enzyme or deep breathing reacts that organic carbon to release from it CO_2. In this sense carbon is cycled. The same can be said of nitrogen as it goes from sluggish N_2 of the atmosphere, via "nitrogen fixatives," to useful N of amino acids. When amino acids released from proteins are converted to nitrogen waste and after conversions become the N_2 of gas in the air it is said that the nitrogen cycle is complete. Elements cycle faster within ecosystems than between but no chemical is entirely isolated. I prefer the idea that Earth is a network of "ecosystems" over any personification of Mother Gaia. My colleague Daniel Botkin would probably define any ecosystem as a set of communities of different species of organisms, living in the same place at the same time, enjoying an influx of external energy and matter. He'd claim, and I agree, that an ecosystem is a volume of Earth surface where organisms recycle energy and matter at a faster rate inside the system than between it and other systems. The material and energy needs of organisms in any ecosystem are met by recycling all of the many chemical compounds required for life maintenance. To "green" Mars, to colonize other planets, or to live for extended periods in space will, of course, require far more than just human settlers and machines. It will require organized, efficient communities. Living together will be as crucial to the colonization of outer space as symbiosis and diversity were to the Paleozoic Era colonization of dry land. Life in space, if it is to occur, will require physical alliances, including new symbioses, among differing life-forms.

New symbioses, forging new patterns of interaction, have already been crucial in the colonization of important parts of Earth. Land dwellers may owe their hold on dry ground to specific symbioses between plants and fungi.

Plant roots and fungi grow together into bumps on roots called mycorrhizae (Fig. 5). Together, fungal-plant complexes settled inhospitable dry regolith: sand, soil, and pebbles.

Life evolved in the sea, but the argument is strong that only interliving—symbiogenesis—made habitation of the hostile new dry land possible for life. Solar ultraviolet radiation, devastating desiccation, and nutrient scarcity were much more serious problems on land 500 million years ago than they are now.

Symbiogenesis developed the Earth's terra firma into occupiable real estate. The early landlocked symbioses were likely not bacterial. The oldest large land organisms that left a fossil record probably were plant-fungal complexes. The oldest plant fossils in the world come from chert, a kind of rock popularly called black flint. The best, most plant-fossiliferous chert comes from a quarry near Rhynie, Scotland. Rhynie fossils are thought to have been preserved in such exquisite detail because of the flow of penetrating water from a nearby silica-rich hot spring. Among the Rhynie chert treasures are fossil chytrids, a kind of protoctist, inside fossilized algae. The algae themselves live inside the stems of 400-million-year-old plants! The quality of the snapshots the fossils give us of the earliest life on land is astounding. One insect preserved intact in Rhynie chert carried in its gut a fungal chlamydospore. (This fancy name refers to a structure resistant to cold and to drying out. Chlamydospores are propagules formed without any sex by the partitioning off of fungal threads.)

The Canadian botanists K. A. Pirozynski and D. W. Malloch propose the idea of "fungal fusion" to help explain the origin of plants 450 million years ago. They hypothesize coevolution of fungi and algae in symbiogenesis: the partners combined. Ultimately plants provided sap for internalized fungi whose mycelial threads developed tough branching and roots. Peter R. Astatt, of the University of California,

Irvine, extends the Pirozynski-Malloch hypothesis, pointing out that plants break down cellulose walls of their cells by using the degradative and absorptive tricks of fungi. Both fungi and plants, for example, secrete chitinase enzymes into the soil. Astatt argues that during their long association with fungi, plants stole and retained fungal genes.

Today mycorrhizae are swollen symbiotic structures, distinct and recognizable. Often colorful, they form symbiogenetically by an interaction of fungi and the root tissue of plants. Mycorrhizae provide the plant partner with mineral nutrients, supplying it with soil phosphorus and nitrogen. The plant supplies the fungal partner with sap, photosynthate food. Mycorrhizal fungi today make chlamydospores strikingly similar to those found in the ancient fossils. Even 450-million-year-old plant remains in the Rhynie chert, including *Rhynia* itself, have swollen roots. Fungi and plants were already locked in productive symbioses at the very beginning of their tenure on dry land.

The move to land was synonymous with the evolution of plants from water-dwelling algae. Survival on land required fortitude: strength, desiccation resistance, and adequate nutrition. Astatt, who has not convinced his colleagues, states that these great discontinuities from the ancestral alga's habitat required symbioses between algae and fungi. Green algae, floating at the ocean's edge, did not just grow big and one day become a plant.

The desert valleys of Antarctica's Victoria Land are an icy hell. Gusts of wind periodically blow over rock and instantaneously freeze the melting ice of summer. Nonetheless, hidden two or three millimeters beneath the rock thrive communities of lichens, a symbiotic mix of fungi, algae, and bacteria that even inhabits porous sandstone. As long as this community can sun itself through the crystalline grains of quartz, it lives. An estimate of the global weight of such fungus-lichen rock dwellers is 13×10^{13} tons, a biomass greater than all life in the ocean! Algae growing under protective cover of fungi cling to sheer rock, extend over its face, and ultimately break it down into soil that can be penetrated by roots of plants and fungal hyphal networks. The hard rock of this spinning planet has been crumbling for hundreds of millions of years into rich, nutritive soil as a result of the fungal-algal partnerships. Lichens, too, are major players in making land habitable for life in temperate climes.

Over billions of years, life, in what the McMenamins call "Hypersea,"[1] extended its domain from its watery home onto dry land. With elegance, novelty, and shocking prodigiousness, life expanded into places it had never gone before. Today the number and diversity of species on land, and species interconnections there, exceed those in the sea, which was life's original habitat. The biomass of life on land is hundreds if not thousands of times greater than the biomass of life in the seas. Much of this massive presence, an estimated 84 percent, is taken up by trees. The foresting of Earth, the dramatic expansion of life beyond its oceanic womb, entailed a dramatic restructuring of the terrestrial environment. For photosynthesizers nutrients such as sulfate and phosphorus that floated freely through the water, the external circulatory system of the oceans, were in short supply on land. These nutrients had to be conducted by the hypersea network itself. The move to land entailed new architecture and infrastructure.

Where life dwelled, water flowed through it. Cytoplasm is more than 80 percent water. Mark McMenamin and his wife, the paleontologist Dianna McMenamin, call attention to profound results of symbiogenetic interconnections by the catchy name "hypersea." What the McMenamins refer to as hypersea is largely the root system of plants that depend on mycorrhizae fungi. Over five thousand kinds of mycorrhizae symbionts, fungi tangled in plant root hairs, can be recognized by name. Vascular plants, of which *Rhynia* is an early example, include all plants except mosses, liverworts, and a few other damp ground covers. Vascular plants have circulatory systems. They are able to pump water from the ground toward their stems and leaves and distribute photosynthate (food) downward. Inconspicuous and underappreciated are their microscopic, mycorrhizal subterranean connections; the literal low profile of the partnerships explains the prodigious success of the plants we see.

Although globally important, mycorrhizal fungi only occasionally attract notice, for example, when they form truffles. These Italian and French delicacies are reproductive spore-forming parts of certain aromatic mycorrhizal fungi attractive to pigs and dogs, which sniff them out of the roots of hardwood trees. Plants with root mycorrhizae are naturally selected: in nutrient-deficient soils they produce heavier seedlings, with greater stores of nitrogen and phosphate, than their counterparts unlinked to fungi. Indeed, 90 percent of living plants have mycorrhizal symbionts. Over 80 percent of all plants perish if deprived of these fungal associates. Hypersea reigns.

The McMenamins' concept needs critical assessment and critical acclaim. The Russian mineralogist Vladimir Vernadsky (1863–1945) recognized life as the great geological force. Anticipating Hypersea, he called living matter "animated water." Animated water is an excellent description of life.[2]

Plants made the move to land by re-creating their wet environment and sealing it within themselves. Trees are prolifically adept at sealing in water, moving it to land, and controlling its evapotranspiration. With their branching networks of tissue strengthened by cellulose and lignin, trees are, of course, vascular plants. Lignin is a complex combination of polyphenolic carbon chemicals that give wood its hardness. The appearance of trees, over 400 million years ago, spurred the entire biosphere upward and outward. The great expansion on land up and out of both sea and fresh water was grounded in the intimacy of plant with fungus, and it still is. Fungi are preeminent in the kingdom of land dwellers. Never photosynthetic, they obtain their food through absorption. They always lack undulipodia so their cells never swim. But wow, can they survive temporary des-iccation! Fungi have patience that far exceeds that of any saint. They sit and wait, and when dampness reappears they take over. Most fungi form intricate mycelial networks, meshes of cytoplasm-filled feeding tubes. Alone, in concert with algae as lichen, or with plant roots as mycorrhizae, they conquered the land and multiplied.

Symbiogenesis was the moon that pulled the tide of life from its oceanic depths to dry land and up into the air. The network of water on land, the animated water of fungi in plants, is the McMenamins' Hypersea. If people ever journey in outer space, the endeavor will never be as machinate and barren as "Star Trek." The vision of sterile engineering eman-cipating us from our planetmates is not only tasteless and boring, it borders on the hideous. No matter how much our own species preoccupies us, life is a far wider system. It is an incredibly complex interdependence of matter and energy among millions of species beyond (and inside) our own skin. These Earth aliens are our relatives, our ancestors, and part of us. They cycle our matter and bring us water and food. Without "the other" we do not survive. Our symbiotic, inter-active, interdependent past is connected through animated waters.

....

CHAPTER 8

GAIA

There is no Silence in the Earth — so silent
As that endured
Which uttered, would discourage Nature
And Haunt the World (1004)

Proprioception, the perception of movement and spatial orientation arising from stimuli inside the body, is a medical concept. Although the name for it is not well known, the phenomenon is familiar to all of us. Our proprioceptors incessantly inform us that we are standing up, inclining our head, squinting our eyes, or clenching our fists. Proprioceptors work as sensory systems not for outside information about others or the environment but inside the body. Nerves attached to muscles fire when they detect motion such as change in positioning of the body. These self-monitoring nerves tell us whether we are standing on our feet or our head or are on the bus at a standstill or jogging along at thirty-five miles an hour. The Earth has enjoyed a proprioceptive system for millennia, since long before humans evolved. Small mammals communicate the coming earthquake or cloudburst. Trees release "volatiles," substances that warn their neighbors that gypsy moth larvae are attacking their leaves. Proprioception, the sensing of self, probably is as old as self itself. I like to think that we people augment and continue to accelerate Gaia's newfangled proprioceptor capability. A fire in the Borneo forest and a crash of a U.S. helicopter in the Italian Alps are broadcast on televised news in New York City. Yet extinct packs of wolves and flocks of dinosaurs enjoyed their own proprioceptive social communication; the global nervous system certainly did not begin with the origin of people. Gaia, the physiologically regulated Earth, enjoyed proprioceptive global communication long before people evolved. The air circulated gas emissions and soluble chemicals from tropical trees, mating-ready insects, and life-threatened bacteria. Love compounds have wafted in spring breezes since the Archean age. But the speed of proprioception has greatly increased with the electronic age.

The second Gaia conference, "Gaia in Oxford," England, April 1996, brought together scientists and environmental activists to discuss superorganisms. Does all life on the planet constitute a superorganism? Is life one single self-regulating entity called Gaia? Does insistence on the superorganism incite comforting but scientifically unwarranted notions of planetary harmony?

These ideas were bandied about by some forty participants, who reached one conclusion: the decision to found *The Geophysiology Society* at East London University. To my delight this decision was reversed in late 1997: Gaia lives; geophysiology is dead. The new organization now calls itself *Gaia: The Society for Research and Education in Earth System Science. Gaia,* the scientific society, was launched on February 9, 1998, at the headquarters of the Royal Society in London. E. O. Wilson, biologist, the world's expert on biodiversity and specialist on ants, their social behavior and technological prowess, sent a videotape of greeting. Holding the formal inauguration in the venue of the oldest and most revered scientific society in the world was a wonderful ploy for Gaia theory. The televised greeting from a well-respected

Harvard professor representing Gaia fans on this side of the Atlantic was also a shrewd move. The increased communication among potential contributors to Gaian science can only raise awareness of the extent of our ignorance about the Earth's surface, on which we so crucially depend.

The Gaia hypothesis is not, as many claim, that "the Earth is a single organism." Yet the Earth, in the biological sense, has a body sustained by complex physiological processes. Life is a planetary-level phenomenon and the Earth has been alive for at least 3,000 million years. To me, the human move to take responsibility for the living Earth is laughable—the rhetoric of the powerless. The planet takes care of us, not we of it. Our self-inflated moral imperative to guide a wayward Earth or heal our sick planet is evidence of our immense capacity for self-delusion. Rather, we need to protect us from ourselves.

The central figure at the 1996 conference was James Lovelock, author of the Gaia hypothesis. Lovelock was the first to claim, in the early 1970s, that the sum of life optimizes the environment for its own use. Biologists rankled at the word *optimizes*. How, they chided, could life plan anything? Lovelock had already thought up the idea of a living Earth in the mid-1960s, years before I met him, when he was a consultant for NASA, helping to devise ways to detect life on Mars. Jim realized that life on any planet would have to use its fluids, which on Earth would be the atmosphere, oceans, lakes, and rivers, to cycle the elements required by life. Nutrients had to be supplied and wastes removed. The chemistry of a living planet, he reasoned, must differ markedly from that of a lifeless one. He realized that even from outer space Earth's atmospheric chemical contradictions are detectable. Our atmosphere contains far too much oxygen in the presence of methane. These gases, highly reactive when mixed, could not coexist at such high concentrations unless the levels were being actively maintained. Other gases in a totally improbable and highly unstable mixture abound. Hydrogen and even nitrogen react explosively in the presence of oxygen, yet they coexist in Earth's atmosphere. Even when Lovelock first wrestled with issues of detecting Martian life, analysis from ground-based telescopes had already shown that Mars, unlike Earth, enjoys a stable atmosphere of nonreactive gases. Lovelock correctly inferred that life could not now exist on Mars. Of course, the *Viking* mission, in its attempt to detect current life, flew to Mars anyway. My opinion is that when the *Viking* lander relayed its data to Earth in 1976, it only confirmed Lovelock's prediction from Gaia theory.

Lovelock's thoughts turned toward Earth. An independent scientist both isolated and free from the academic mainstream, Jim went about pursuing his interests in his own way. He is a prolific inventor; his major contribution is the electron capture device, a detector attached to an instrument called a gas chromatograph, used to measure concentrations of certain reactive gases in the air such as chlorofluorocarbons. Lovelock's equipment was redesigned and widely sold by Hewlett-Packard. The devise allowed Sherwood Rowland and Mario Molina at MIT to receive the 1995 Nobel prize in chemistry for showing how spray-can and other gases react to destroy the stratospheric ozone layer. Jim's other accomplishments include vindicating Rachel Carson's claims of the pervasive impact of pesticides, which she brought to the nation's attention in *Silent Spring*. While working for the Medical Research Council (MRC) in Britain on cryogenics, cold-temperature biology, Jim developed a means to freeze and thaw animals and their sperm. He thawed frozen specimens in a chamber he constructed, a microwave oven of sorts (he did not, however, patent this invention; patenting, a long and costly process, is confusing and highly social, exactly the kind of activity Lovelock avoids).

When Jim became interested in how life affected Earth's atmosphere, he pursued his science without institutional help or grant money. He simply plunged in and measured gases at his own expense. He never ceased communication with colleagues and students. The result of his incessant activity was the development of Gaia theory.

We began to correspond in the early 1970s. In one of his first answers to my letters, Jim wrote that he was troubled by methane. Why is this gas, which reacts so strongly with oxygen, always measurably present in Earth's atmosphere? It should disappear. Suspecting life from the beginning, he asked whether I knew what could possibly produce this gas. I responded the way anyone who reads microbiology would. Methane gas is produced by bacteria, mainly the methanogens that live in waterlogged soil or in cattle rumen. The metabolic product of methanogenic bacteria is released in copious quantities not in cow flatulence (as I always thought) but in their belches. Methane is released into the air through the mouths of calves, bulls, and cows. Atmospheric methane quickly reacts with oxygen to produce carbon dioxide. Clearly air methane is replenished on a regular basis because it is always present at concentrations from two to seven parts per million. Lovelock realized that atmospheric methane concentrations must therefore be regulated by life. Other examples of gas regulation were likely.

Geological clues suggest our planet has become cooler in the past three billion years. Astronomers insist that the sun, a typical star, has become brighter. The sun should have increasingly heated Earth's surface over its immense past. Temperature and atmospheric regulation, Jim reasoned, must occur on a global scale. As he realized that these vital environmental conditions must be actively controlled, Lovelock was led to propose that life maintains its environment.

Borrowing a term from physiology, Lovelock pointed out that our planetary environment is homeostatic. Just as our bodies, like those of all mammals, maintain a relatively stable internal temperature despite changing conditions, the

Earth system keeps its temperature and atmospheric composition stable. In engineering terms, Lovelock wrote, atmospheric temperature is regulated around given set points by
negative feedback. His claims that "life sets environmental
temperature at an optimum" were misunderstood—criticized
or, more frequently, ignored. Lovelock increasingly thought
of this planetary regulatory system as central to understanding life on Earth.[1]

The term *Gaia* was suggested to Lovelock by the novelist
William Golding, author of *Lord of the Flies.* In the early
1970s, they both lived in Bowerchalke, Wiltshire, England.
Lovelock asked his neighbor whether he could replace the
cumbersome phrase "a cybernetic system with homeostatic
tendencies as detected by chemical anomalies in the Earth's
atmosphere" with a term meaning "Earth." "I need a good
four-letter word," he said. On walks around the countryside
in that gorgeous part of southern England near the chalk
downs, Golding suggested Gaia. The ancient Greek word for
"Mother Earth," *Gaia* provides an etymological root of many
scientific terms, such as *geo*logy, *geo*metry, and *Pan*gaea.

The name caught on all too well. Environmentalists and
religiously inclined people, attracted to the idea of a native
goddess with power, latched onto it, giving Gaia a distinctly
nonscientific connotation. Just before Oxford '96 Jim proposed the term *geophysiology* for the study of the planetary
surface as an organismlike body in which geology and biology are "tightly coupled," that is, intimately linked.

Many scientists are still hostile to Gaia, both the word
and the idea, perhaps because it is so resonant with anti-
science and anti-intellectual folks. In popular culture, inso-
far as the term is at all familiar, it refers to the notion of
Mother Earth as a single organism. Gaia, a living goddess
beyond human knowledge, will supposedly punish or
reward us for our environmental insults or blessings to her
body. I regret this personification.

As detailed in Jim's theory about the planetary system, Gaia is not an organism. Any organism must either eat or, by photosynthesis or chemosynthesis, produce its own food. All organisms produce waste. The second law of thermodynamics speaks clearly on this score: to maintain a body's organization energy must be lost, dissipated as heat. No organism feeds on its own waste. Gaia, the living Earth, far transcends any single organism or even any population. One organism's waste is another's food. Failing to distinguish anyone's food from someone else's waste, the Gaian system recycles matter on the global level. Gaia, the system, emerges from ten million or more connected living species that form its incessantly active body. Far from being fragile or consciously petulant, planetary life is highly resilient. As they unwittingly obey the second law of thermodynamics, all beings seek energy and food sources. All produce useless heat and chemical waste. This is their biological imperative. Each grows and, as it does, it pressures many others around it. The sum of planetary life, Gaia, displays a physiology that we recognize as environmental regulation. Gaia itself is not an organism directly selected among many. It is an emergent property of interaction among organisms, the spherical planet on which they reside, and an energy source, the sun. Furthermore Gaia is an ancient phenomenon. Trillions of jostling, feeding, mating, exuding beings compose her planetary system. Gaia, a tough bitch, is not at all threatened by humans. Planetary life survived at least three billion years before humanity was even the dream of a lively ape with a yearning for a hairless mate.

We need honesty. We need to be freed from our species-specific arrogance. No evidence exists that we are "chosen," the unique species for which all the others were made. Nor are we the most important one because we are so numerous, powerful, and dangerous. Our tenacious illusion of special dispensation belies our true status as upright mammalian weeds.

In popular culture the confused idea of Gaia strikes mythological chords. Gaia resonates with our longing for significance in our short Earth-bound lives. Misstated Gaia supports latter-day Puritanism: feminist discourse on the dangers of "rape" and destruction of the sunlit Earth. We have for centuries personified nature. The co-optation of Gaia theory by science-haters and media-mongers is striking. The former blame science, only a way of knowing, for the excesses of technology, and the latter use science to justify their crass television and magazine salesmanship. Popularized, exaggerated, or maligned Gaia theory does not just mean nature conservation and a return to the Goddess. Gaia is the regulated surface of the planet incessantly creating new environments and new organisms. But the planet is not human, nor does it belong to humans. No human culture, despite its inventiveness, can kill life on this planet, were it even to try. More an enormous collection of interacting ecosystems, the Earth as Gaian regulatory physiology transcends all individual organisms. Humans are not the center of life, nor is any other single species. Humans are not even central to life. We are a recent, rapidly growing part of an enormous ancient whole.

Gaia is neither vicious nor nurturing in relation to humanity; it is a convenient name for an Earthwide phenomenon: temperature, acidity/alkalinity, and gas composition regulation. Gaia is the series of interacting ecosystems that compose a single huge ecosystem at the Earth's surface. Period.

Fossil evidence records that Earth life in its 3,000-million-year history has withstood numerous impacts equal to or greater than the total detonation of all five thousand stockpiled nuclear bombs. Life, especially bacterial life, is resilient. It has fed on disaster and destruction from the beginning. Gaia incorporates the ecological crises of her components, responds brilliantly, and new necessity becomes the mother of invention.

Bacteria at first removed the hydrogen (H_2) they needed for their bodies directly from the air. Later they took up the hydrogen sulfide (H_2S) belched up from volcanoes. Eventually, blue-green bacteria wrenched hydrogen atoms from water (H_2O). Oxygen was expelled as a metabolic waste product. This waste, at first disastrous, eventually powered life's continued growth. New wastes test life's tolerance and stimulate life's creativity. The oxygen we need to breathe began as a toxin; it still is. The oxygen release from millions of cyanobacteria resulted in a holocaust far more profound than any human environmental activity. Pollution is natural. "Waste not" is a exhortation, not a description. The cyanobacteria's waste became our fresh air. We humans obtain the hydrogen we require by eating plants or other animals. We cannot do without. Often newly evolved beings grow and expand rapidly by exploiting the energy, food supplies, or wastes of others. But population expansion always ceases because none can eat or breathe its own waste. Populations crash or slow down as they meet impediments to expansion. This check on growth is precisely what Charles Darwin meant by "natural selection." Gaia is the sum of these growing, interacting, and dying populations; a multi-species planetary covering, composed of myriad very different beings, Gaia is the only giant ecosystem on Earth.

Unlike any of its component ecosystems, Gaia is the genius of recycling. Roughly one-fifth of Earth's atmosphere is oxygen (O_2). Combined with hydrogen (H_2) or hydrogen-containing gases (CH_4, H_2S, NH_3), oxygen causes explosions and fires. Reactions that release energy change the reactive gases into their "spent" or less reactive by-products. Reactive gases such as hydrogen (H_2), methane (CH_4), ammonia (NH_3), methyl iodide (CH_3I), methyl chloride (CH_3Cl), and various sulfur gases are detectable in Earth's atmosphere because they are continually produced by waste-producing life faster than they can react.

My longtime colleague and former student Lorraine Olendzenski and I made a video at the University of Massachusetts, Amherst (formerly the Massachusetts College of Agriculture). In the video our wonderful friend Betsy Blunt Harris, a microbiologist who has been teaching the teachers of microbiology labs for over a decade, reaches her gloved hand through a fashioned opening, a "fistula," in the side of a healthy cow. Betsy's fingers contact the cow's rumen, a special huge stomach, one of the four found in all cows and their ruminant relatives. She removes brown fibrous mush, mostly partially digested grass, from the fistula. The mush is so packed with microbes that it must be highly diluted before we view it with a microscope. The cow's microbial community includes weird swimming cells, ciliate protists. Many bacteria, most of them smaller than the ciliates, also inhabit rumens. These microbes do the work of grass digestion.[2] Without them no cow digests the cellulose of grass. Indeed, the cellulose-degrading microbes, in a very real sense, are the cow. Without them the cow would not swallow, ferment, regurgitate, and reswallow. No cow would be grass-eating or cud-chewing without the microbial middlemen. One of the gaseous products of grass digestion is methane. Cows belch huge quantities of it. Bovine methane is part of the reason that Earth's air is a highly unstable chemical mixture. Wood-eating termites also release methane. Like cows, they harbor gut microbes that break down cellulose into various chemical products. Carbon dioxide, methane, nitrogen, and sulfur-containing gases are expelled into the air through the anuses of millions of termites. The long-term unstable gas systems of the atmosphere result from incessant microbial life.

Generalizing these findings, Lovelock proposed that the entire planetary air system is "metastable," stable in its reactive instability. The persistence of chemical reactivity arises from the combined actions of living beings. The entire planetary surface, not just the living bodies but the atmosphere that we think of as an inert background, is so far from chemical equilibrium that the entire planetary surface is best regarded as alive.

I cannot stress strongly enough that Gaia is not a single organism. My Gaia is no vague, quaint notion of a mother Earth who nurtures us. The Gaia hypothesis is science.[3] The surface of the planet, Gaia theory posits, behaves as a physiological system in certain limited ways. The aspects that are physiologically controlled include surface temperature, atmospheric composition of reactive gases, including oxygen, and pH or acidity-alkalinity.

I suspect scientists will seek Gaian explanations for many phenomena, such as the alteration of wet and dry climate cycles and the current distribution of gold, iron, phosphates, and other minerals. Gaia, meaning a body with a controlled physiology in the celestial-planetary and biological sense, is the only name that can both unite an unruly group of scientists and make their work accessible to the international public. Just as the human body is sharply bounded by skin, temperature differences, blood chemistry, and a calcium phosphate skeleton, so is Earth distinguished from its surroundings by its persistently anomalous atmosphere, its steady temperature, and its unusual limestone and granitic rocks. Lovelock compares the chemistry of Earth's atmosphere to a sand castle found on a beach, or to a bird's nest. They, too, are obvious products of life. The planet's surface is not just physical, geological, and chemical, or even just geochemical. Rather, it is geophysiological: it displays the attributes of a living body composed of the aggregate of Earth's incessantly interactive life.

Physiological chemistry, what we call metabolism, results from the activity of living beings. Just how closely linked the chemical systems of Gaia are remains a matter of debate. "Weak Gaia" holds that the environment and life are coupled; they coevolve. Few disagree. Many scientists dismiss this idea as old news. "Strong Gaia" states that the planet with its life, a single living system, is regulated in certain aspects by that life. This is the idea that elicits the derision of certain biologists, especially those who call themselves neo-Darwinists. Led by Richard Dawkins of Oxford, these scientists reject the idea of a unified planetary system that has not evolved through natural selection with other planetary systems. Lovelock, who has been accused of wavering, claims never to have abandoned "Strong Gaia," the term coined by J. Kirshner, a scientist-philosopher at the University of California, Berkeley. At the 1988 Chapman Conference of the American Geophysical Union meeting, Kirshner ridiculed "Strong Gaia" just after he had snidely defended it. You can read about Kirshner's and everyone else's strong response to Gaia and philosophy in the meeting report edited by Schneider and Boston.[4] Lovelock admits, though, that he gave up his original notion that Gaia is "teleological." He no longer asserts that the living planetary system behaves together to optimize conditions for all its members. Biodiversity is an absolute requirement for Gaian persistence. There is no most-favored-species list. Any organism does its thing: grows and attempts to reproduce. Selection pressures, the insistence of all organisms that they grow and reproduce, favor certain types of life under certain specified conditions. These grow, expand, remove waste, and recycle. As they do, they place enormous selection pressures on still different types of life. The result is Gaia. Were there no life, temperature and gas composition would be predictable solely from physical factors. The sun's output of energy and the rules of chemistry and physics would determine Earth's surface properties. But these properties deviate significantly from predictions based on physics and chemistry alone. The non-biological sciences do not suffice to explain the Earth's surface environment. When the multifaceted roles of gas-producing, temperature-altering living organisms are taken into account, the disparities disappear. Gaia theory is useful science.

Any new idea generates criticism, especially in science, where criticism is institutionalized by peer review publication and repeatable experiment. The Gaia idea requires geologists, geochemists, atmospheric chemists, and even meteorologists to understand science outside their own fields. They must study biology, especially microbiology. But academic apartheid breeds resistance. Accepting Gaia would lead to action that people in related fields are loath to take.

There is nothing new in Gaia but the name, claim some critics. Others assert that the proposition that the Earth's surface is alive is so broad that it cannot be tested. Not necessarily: if we define life as a reproducing system capable of natural selection, then Gaia is living. The easiest way to see this is through a simple thought experiment. Imagine that a spacecraft carrying microbes, fungi, animals, and plants is sent to Mars. Let it produce its own food and cycle its waste, and let it persist for two hundred years. Gaia is the recycling system of life as a whole. A budding off of one Gaia to produce two would have occurred. The construction of such a miniature Gaia would represent de facto reproduction. Dorion Sagan's book *Biosphere* makes this case clearly.[5]

Another criticism of Gaia theory speaks to scientists' fears. That Gaia theory resonates with ancient beliefs of Mother Earth seems to some critics to make it dangerously unscientific. These critics claim no planetary entity can act in a concerted way if it lacks conscious control. How does the planet know when to raise or lower atmospheric oxygen concentration to keep it at about 20 percent? This level hovers between a global fire hazard and the risk of widespread death by asphyxiation. How can Gaia "arrange" to remove salt from oceans to save its inhabitants from a threatening level of salinity? How can "she" cool her entire body to com-

pensate for the increasing luminosity of the sun? How does
Gaia know to regulate ocean cloud cover in service to temperature? Just who is this Gaia?

Lovelock replies that Gaia requires no consciousness to
adjust to the planetary environment. Recent work in mathematics, called fractal geometry, shows that elaborate graphics
can be made not by an artist with a finished idea but by repe-
titions of simple computer steps called algorithms. Life produces fascinating "designs" in a similar way by repeating the
chemical cycles of its cellular growth and reproduction.
Order is generated by nonconscious repetitious activities.
Gaia, as the interweaving network of all life, is alive, aware,
and conscious to various degrees in all its cells, bodies, and
societies. Analogous to proprioception, Gaian patterns
appear to be planned but occur in the absence of any central
"head" or "brain." Proprioception, as self-awareness,
evolved long before animals evolved, and long before their
brains did. Sensitivity, awareness, and responses of plants,
protocists, fungi, bacteria, and animals, each in its local
environment, constitute the repeating pattern that ultimately
underlies global sensitivity and the response of Gaia "herself." With his colleague and former Ph.D. student Andy
Watson, Lovelock developed a computer model called
"Daisyworld." They assume a planet, for example, on which
live only white and black daisies. The planet is exposed to
the radiation of a star, modeled after our sun, that grows in
luminosity over millions of years. Without any extraneous
assumptions, without sex or evolution, without mystical pre-
suppositions of planetary consciousness, the daisies of
Daisyworld cool their world despite the warming sun.

The assumptions are straightforward. Black daisies tend
to absorb heat and white daisies to reflect it. Neither flower
grows below 10 degrees Celsius and they all die above 45
degrees Celsius. Within this range, black daisies tend to
absorb local heat and therefore grow faster in colder conditions. White daisies, since they reflect and lose more heat in

warmer conditions, thrive to produce more offspring. Let us begin with the black daisy world. As the sun increases in luminosity, the black daisies grow, expanding their surface area, absorbing heat, and heating up their surroundings. As the black daisies heat up more of the surrounding land surface, the surface itself warms, permitting even more population growth. This positive feedback continues until daisy growth has so heated the surroundings that white daisies begin to crowd out the black ones. Being less absorbent and more reflective, the white daisies begin to cool down the planet. The cumulative result of these actions is to heat the planetary surface when it is cooler during the early evolution of the sun, and then to keep the planet relatively cool as solar luminosity increases. Despite the ever-hotter sun, the planet maintains a long plateau of stable temperature.

Daisyworld proved to be a turning point in Gaian science. Stephan Harding, professor at Schumacher College in Devon, England, now models Daisyworlds with twenty-three different-colored species of daisies as well as herbivores that eat the daisies and carnivores that eat the herbivores. No relationship emerges in these models between what is good for a particular species and what is good for the planet as a whole. The population growth of one kind of organism may lead to its own collapse. What has emerged is the mathematical outline of an overlap between natural selection and global temperature regulation. Global temperature regulation is a paradigmatic example of Gaian behavior. Harding's models indicate differential survival behaves to buttress or even spawn global-level consequences. Biologists are less reluctant to embrace Gaia theory. Temperature regulation is a physiological function not only of Daisyworld but of the bodies and the societies of life. Mammals, tuna, skunk cabbage plants, and beehives all regulate their temperatures to within a few degrees. How do plant cells or hive-dwelling bees "know" how to maintain temperature? Whatever the answer in principle, the tuna, skunk cabbage, bees, and mouse cells display the same sort of physiological regulation that prevails across the planet.

Gaia, in all her symbiogenetic glory, is inherently expansive, subtle, aesthetic, ancient, and exquisitely resilient. No planetoid collisions or nuclear explosions have ever threatened Gaia as a whole. So far the only way in which we humans prove our dominance is by expansion. We remain brazen, crass, and recent, even as we become more numerous. Our toughness is a delusion. Have we the intelligence and discipline to resist our tendency to grow without limit? The planet will not permit our populations to continue to expand. Runaway populations of bacteria, locusts, roaches, mice, and grass always collapse. Their own wastes disgust as crowding and severe shortage ensue. Diseases, after opportunistic expanding populations of the "other," follow, taking their cue from destructive behavior and social disintegration. Even herbivores, if desperate, become vicious predators and cannibals. Cows will hunt rabbits or eat their calves, and many mammals will vie for the meat of their runted littermates. Population overgrowth leads to stress, and stress depresses population overgrowth—an example of a Gaian-regulated cycle.

We people are just like our planetmates. We cannot put an end to nature; we can only pose a threat to ourselves. The notion that we can destroy all life, including bacteria thriving in the water tanks of nuclear power plants or boiling hot vents, is ludicrous. I hear our nonhuman brethren snickering: "Got along without you before I met you, gonna get along without you now," they sing about us in harmony. Most of them, the microbes, the whales, the insects, the seed plants, and the birds, are still singing. The tropical forest trees are humming to themselves, waiting for us to finish our arrogant logging so they can get back to their business of growth as usual. And they will continue their cacophonies and harmonies long after we are gone.

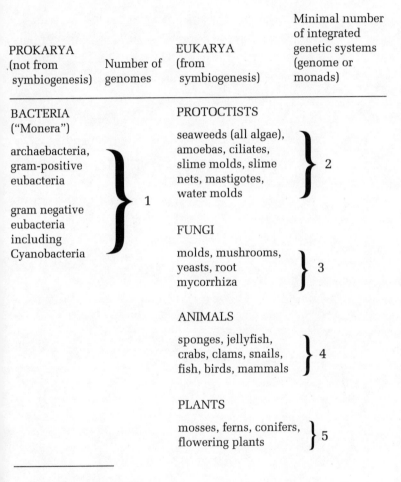

PROKARYA (not from symbiogenesis)	Number of genomes	EUKARYA (from symbiogenesis)	Minimal number of integrated genetic systems (genome or monads)
BACTERIA ("Monera")		PROTOCTISTS	
archaebacteria, gram-positive eubacteria	} 1	seaweeds (all algae), amoebas, ciliates, slime molds, slime nets, mastigotes, water molds	} 2
gram negative eubacteria including Cyanobacteria		FUNGI	
		molds, mushrooms, yeasts, root mycorrhiza	} 3
		ANIMALS	
		sponges, jellyfish, crabs, clams, snails, fish, birds, mammals	} 4
		PLANTS	
		mosses, ferns, conifers, flowering plants	} 5

---

* See Figure 2, page 31.

PROLOGUE

1. Lynn Margulis and Dorion Sagan, *Slanted Truths: Essays on Gaia, Symbiosis, and Evolution* (New York: Copernicus Books, 1997): "Sunday with J. Robert Oppenheimer" and other works, many previously published, describe the "big trouble" with neo-Darwinian biology and the generally destructive effects of academic apartheid. Essays detail the context of life as a planetary phenomenon in which symbiogenesis, especially via bacterial associations, is the major source of evolutionary novelty.

   Lynn Margulis and Dorion Sagan, *What Is Sex?* (New York: Simon & Schuster, 1997): a philosophical and pictorial inquiry into the evolution of sex from its philosophical beginnings in an energetic universe to cybersex and beyond.

   Lynn Margulis and Dorion Sagan, *What Is Life?* (New York: Simon & Schuster, 1996): illustrated with color photographs and black-and-white drawings, this is a philosophical and scientific exploration of one of history's most fascinating questions. It argues for a concept of life that transcends both mechanism and vitalism. Including sections on the solar basis of the global economy and humanity's status as superorganism, it emphasizes the neglected role of free will in evolutionary development.

   Gaia to Micrososm, Vol. 1, *Planetary Life* (Dubuque: Kendall/Hunt, 1994): four short videos: *From Bacteria to Biosphere*; *Photosynthetic Bacteria—Red Sunlight Transformers*; Spirosymplokos deltaeiberi—*Microbial Mats and Mud Puddles; and* Ophrydium versatile: *What Is an Individual?*

Dorion Sagan and Lynn Margulis, *Garden of Microbial Delights: A Practical Guide to the Subvisible World* (Dubuque: Kendall/Hunt, 1993): a guide to the history of knowledge, diversity, and usefulness of the microscopic world, including notes on how to keep microbial pets. Written with teachers and students, nature lovers, and natural history museum goers in mind, it is abundantly illustrated.

Lynn Margulis, *Five Kingdoms Poster,* illustrated by Christie Lyons, designed by Dorion Sagan (Rochester, N.Y.: Ward's, 1992): set includes a teacher's guide to the poster and a guide to classroom activities.

Lynn Margulis and Dorion Sagan, *Origins of Sex: Three Billion Years of Genetic Recombination* (New Haven, Conn.: Yale University Press, 1991): the description and time of appearance of DNA recombination in bacteria, cell fusion in protists, appearance of gender, alternation of generations, and other sexual processes in the context of their evolution.

Lynn Margulis and Dorion Sagan, *Mystery Dance: On the Evolution of Human Sexuality* (New York: Summit Books, 1991): summary of the multilevel effects of ancestors on human sexual form and behavior. Lacanian psychoanalysis, reptilian origins of the mammal brain, sperm competition, and the role of jealous violence in genetic propagation figure among the discussions. The story begins with the most recently evolved ancestors and ends with the origins of meiosis and bacterial sex in unicells, irradiated by ultraviolet light.

Lynn Margulis and Dorion Sagan, *Microcosmos: Four Billion Years of Evolution from Our Microbial Ancestors,* foreword by Dr. Lewis Thomas (Berkeley: University of California Press, paper ed., 1998): popular account of early life, including origin of nucleated cells, the originally toxic buildup of oxygen in Earth's atmosphere, and the appearance of plants and animals from colonies of microbes. A new author's preface recalls the need to move beyond the book's strategic inversion of microbes over humans. The analysis begins with the origin of the universe, 15 billion years ago, and ends with speculation on the future of life, with and without humans.

2. Lois Byrnes, Deep Time Associates, P.O. Box 58, Rockport, Massachusetts, created the exhibit "What Is Life?" at the New England Science Center, 222 Harrington Way, Worcester, Massachusetts. Displaying the art of Christie Lyons, our videos, and other modes of communicating SET and Gaia, this and other excellent expositions are open to the public Tuesday through Saturday, 10:00 A.M.–5:00 P.M.

3. Our relationship with nature and God, profound and complex, is gorgeously analyzed by Evan Eisenberg in his new work, *Ecology of Eden* (New York: Alfred Knopf, 1998).

CHAPTER ONE
....

1. I. E. Wallin, *Symbioticism and the Origin of the Species* (Baltimore: Williams & Wilkins, 1927). Charles Darwin, *On the Origin of the Species by Means of Natural Selection or the Preservation of Favored Races in the Struggle for Life* (London: Murray, 1859).
2. Sorin Sonea and Maurice Panisset, *A New Bacteriology* (Sudbury Mass.: Jones & Bartlett Publishers, 1993).
3. Paul Nardon and A.M. Grenier, "Serial Endosymbiosis Theory and Weevil Evolution: The Role of Symbiosis," in L. Margulis and R. Fester, eds., *Symbiosis as a Source of Evolutionary Innovation* (Cambridge, Mass.: MIT Press, 1991).
4. These organisms and their activity are strikingly illustrated in videos: see Sciencewriters, *Gaia to Microorganism* (Dubuque, Iowa: Kendall/Hunt Publishing Company, 1996), and Lorraine Olendzenski, Lynn Margulis, and Steve Goodwin, *Looking at Microbes: The Microbiology Laboratory for Students* (Sudbury, Mass.: Jones and Bartlett Publishers, 1998).

CHAPTER TWO
....

1. G. G. Simpson, *An Autobiography* (New York: Columbia University Press, 1977).
2. For a technical description of the heredity, origin, and evolution of these organelles see the second edition of *Symbiosis in Cell Evolution* (New York: W. H. Freeman, 1993). For a less technical but certainly adequate view of the same material see Lynn Margulis and Dorion Sagan, *Microcosmos: Four Billion Years of Evolution from Our Microbial Ancestors*. For the engaging and remarkable history of cytoplasmic genetics see J. Sapp, *Beyond the Gene* (New York: Cambridge University Press, 1987).
3. J. Sapp, *Evolution by Association: A History of Symbiosis* (New York: Oxford University Press, 1994).
4. B. Ephrussi, *Nucleo-cytoplasmic Relations in Micro-Organisms* (Oxford, U.K.: Clarendon Press, 1953).

5. See Sapp's *Beyond the Gene,* note 2 above.

6. J. Sapp, *Evolution by Association: A History of Symbiosis* (New York: Oxford University Press, 1994). Dobzhansky: "Nothing in biology makes sense except in the light of evolution," 1973 quote, p. 187.

7. R. Sager and F. Ryan, *Cell Heredity* (New York: John Wiley & Sons, 1961); also see G. Pontecorvo, *Trends in Genetic Analysis* (New York: Columbia University Press, 1959).

8. E. B. Wilson, *The Cell in Development and Heredity,* 3d ed. (New York: Macmillan Co., 1928).

9. L. N. Khakhina, *Concepts of Symbiogenesis: A Historical and Critical Study of the Research of the Russian Botanists* (New Haven, Conn.: Yale University Press, 1992).

CHAPTER THREE
....

1. D. C. Smith, "From Extracellular to Intracellular: The Establishment of a Symbiosis," in *The Cell as a Habitat,* vol. 204 (London: The Royal Society, 1979) pp. 115–130.

CHAPTER FOUR
....

1. A concept of the intrepid and prolific nineteenth-century German explorer and biologist Christian Ehrenberg. See *What Is Life?* in note 1, p. 131 above.

2. H. F. Copeland, *Classification of the Lower Organisms* (Palo Alto, Calif.: Palo Alto Books, 1956).

3. R. H. Whittaker, *Community Ecology,* 2nd ed. (New York: Macmillan, 1975).

4. R. H. Whittaker, "New Concepts of Kingdoms," *Science* 163: 150–160, 1969.

5. Microbes include bacteria, the smaller protoctists, and the smaller fungi. See Lynn Margulis and Karlene V. Schwartz, *Five Kingdoms: An Illustrated Guide to the Phyla of Life on Earth,* 3d ed. (New York: W. H. Freeman, 1998). A description of each phylum of bacteria, protoctists, fungi, animals, and plants (approximately one hundred phyla) is accompanied by a photograph of an exemplary organism, alive if possible, and a labeled drawing of a typical phylum member.

6. The protoctists are drawn in their many natural habitats, from desert to open ocean in L. Margulis, K. V. Schwartz and M. Dolan. *Diversity of Life on Earth: Illustrated Five Kingdoms* (Sudbury, Mass.: Jones and Bartlett 1999). Larger more familiar organisms accompany microbes in all the drawings.

CHAPTER FIVE
....

1. J. Morowitz, *Mayonnaise and the Origin of Life: Thoughts of Minds and Molecules* (Woodbridge, Conn.: Ox Bow Press, 1985).
2. D. Deamer and G. Fleischaker, *Origins of Life: The Central Concepts* (Sudbury, Mass.: Jones & Bartlett, 1994).
3. Francis Crick, *Life Itself: Its Origins and Nature* (New York: Simon & Schuster, 1981).
4. H. J. Morowitz, *Beginning of Cellular Life* (New Haven, Conn.: Yale University Press, 1992).
5. F. Dyson, *Origin of Life* (Cambridge: Cambridge University Press, 1987).
6. M. T. Madigan, J. M. Matinko, J. Parker, *Brock Biology of Microorganisms*, 8th edition, (Upper Saddle River, New Jersey: Prentice-Hall, 1997).

CHAPTER SIX
....

1. Lynda J. Goff, "Symbiosis, Interspecific Gene Transfer and the Evolution of New Species: A Case Study in the Parasitic Red Algae," in L. Margulis and R. Fester, eds., *Symbiosis as a Source of Evolutionary Innovation* (Cambridge, Mass.: MIT Press, 1991).
2. Lynn Margulis and Dorion Sagan, *What Is Sex?* (New York: Simon & Schuster, 1997) and *Origins of Sex: Three Billion Years of Genetic Recombination* (New Haven, Conn.: Yale University Press, 1991).
3. L. R. Cleveland, "The Origin and Evolution of Meiosis," *Science*, volume 105, pages 287–288, 1947.

CHAPTER SEVEN
....

1. M. A. McMenamin and D. S. McMenamin, *Hypersea: Life on Land,* (New York: Columbia University Press, 1994).

2. V. I. Vernadsky, *The Biosphere* (New York: Copernicus, Springer-Verlag, 1998; 1926 in Russian).

CHAPTER EIGHT

1. James E. Lovelock, *Gaia: A New Look at Life on Earth.* (Oxford, U. K.: Oxford University Press, 1979).
2. Lorraine Olendzenski, Lynn Margulis, and Steve Goodwin, *Looking at Microbes* Videos vol. 1, *The Microbiology Laboratory for Students,* vol 2. *Microbe's World* (Sudbury, Mass.: Jones and Barlett Publishers, 1998).
3. P. Bunyard, ed., *Gaia in Action: Science of the Living Earth* (Edinburgh, U. K.: Floris Books, 1996)
4. S. Schneider and P. Boston, *Scientists on Gaia* (Cambridge, Mass.: MIT Press, 1990).
5. Ibid.

INDEX

Abortive
 cannibalism, 89,
 99–103
Acidity/alkalinity,
 120, 123
Alanine, 76
Algae, 33–34, 55, 61,
 62, 89, 93
 Chlamydomonas,
 28, 47, 97–98
 Chlorella, 10, 11
 chloroplasts, 19, 22,
 29, 36–38, 40–41,
 53
 coevolution with
 fungi, 107–109
 Convoluta
 roscoffensis and,
 9–10
 evolution of plants
 and, 95–98,
 108–111
 origins of, 34–37
Altman, Sidney, 82
Amino acids, 43–44,
 76–77, 106. *See*
 also Proteins
Ammonia, 83, 121
Amoeba, 8, 60, 62
Anaerobes, 35

Anastomosis, 52
Animals, 52–53, 54,
 57, 68
 evolution of, 93–95
 oldest fossils of, 93
 plant-animals, 9–10
 sex among, 88–89,
 99
Animated water, 110
Antarctica, 108–109
Anthropocentrism,
 3–4
Archaea, 65–66
Archaebacteria,
 34–36, 38, 39,
 42–45, 47, 67,
 102–103
Aristotle, 56, 57–58
Astatt, Peter R.,
 107–108
Autocatalytic
 systems, 78
Autotrophs, 83–84
Axons, 48

Bacteria
 abortive cannibal-
 ism among, 89,
 99–103

as basis of
 evolution, 4,
 10–11, 53, 56, 62,
 64–65, 68, 70, 121
as basis of serial
 encosymbiosis
 theory (SET),
 30–32, 34–49
double inheritance
 systems of, 25–26
as enemy agents,
 55–56, 75
as former organelles,
 34
in fruit fly research,
 8
incorporation inside
 plants and
 animals, 6, 25
lack of species in, 6
nitrogen fixing, 5,
 106
nonnuclear cell
 parts as remnants
 of, 25
and origins of life
 problem, 70–85
outside nucleus but
 inside cells, 28

Bacteria (*cont.*)
  photosynthetic,
    9–10, 22, 36–37,
    55, 109
  sexual habits of,
    87–89
Bateson, Gregory, 51
Bateson, Mary
  Catherine, 18
Beisson, Jannine, 27
Belousov-
  Zhabotinsky
  system, 78–79
Bestiaries, 58
Big Bang, 72
Biochemical
  cytology, 21
Biodiversity, 56, 114
Biomass, 108–109
Biophysical cytology,
  21
Biosphere (D. Sagan),
  125
Boston, P., 124
Botkin, Daniel, 106
Branches, 52, 68
Byrnes, Lois, 11

Calcium, in animal
  evolution, 94
Cambrian, 93–95
Cannibalism,
  abortive, 89,
    99–103
Carbon dioxide, 83,
    105–106, 117, 122
Carson, Rachel, 116
Cavalier-Smith, Tom,
    39, 48
Cech, Thomas, 82
Cell genetics

cytoplasmic. *See*
  Cytoplasmic
  genetics
nature of cells and,
    33–34, 69
nuclear. *See* Nuclear
  genetics
origins of life
  theories and,
    79–85
*Cell in Development
  and Heredity, The*
  (Wilson), 25
Cell sex, 98
Cellular interliving,
    20
Centriole-
  kinetosomes,
    41–49
  Henneguy-
    Lenhossek theory
    of, 46
  microtubules of, 48
  nature of, 41
  as seeds, 42–43
Chatton, Edouard, 28
Chemical clocks, 78
Chemistry,
  connection of
  genetics with,
    24–25
Chemotrophs, 83–84
Chert, 107, 108
Chlamydomonas, 28,
    47, 97–98
Chlamydospores,
    107, 108
Chlorella, 10, 11
Chloroplasts, 19, 22,
    29, 36–38
  bacterial origins of,
    53

direct filiation
  theory of origins
  of, 40–41
Chromatin, 36
Chromosomes, 24
  as basis of heredity,
    21
  chromatin and, 36
  defined, 21
  extra, 101–102
  genes located on, 21,
    22
Ciliate genetics,
    27–29
  direct filiation
    theory of origins
    of, 40–41
  spirochete
    hypothesis for
    origins of cilia,
    39–49
Clams, 10
Classes, 59
Cleveland, Lemuel
  Roscoe, 28,
    99–103
Cloning, 93, 97
Cloud, Preston, 94
Coefficients of
  selection, 18
Continuity of the
  germ plasm, 23
*Convoluta
  roscoffensis*, 9–10
Copeland, Herbert P.,
    60–61, 62
Correns, C., 22
Crick, Francis, 73–74
Crow, James F.,
    17–18
Cuvier, Georges, 59

Cyanobacteria, 37, 38, 40, 83, 121
Cytoplasm, 37, 109
Cytoplasmic factors, 22, 24–25
Cytoplasmic genetics
basic concepts of, 19, 20–22, 28
ciliate genetics in, 27–29
heredity in, 27
origins of research in, 19, 20–22

Daisyworld, 126–128
Danielli, James F., 29
Darwin, Charles, 4, 6, 20, 23, 59–60, 121
Dawkins, Richard, 124
Death
abortive cannibalism, 89, 99–103
programmed, 90
deBary, Anton, 33
Dendrites, 48
DeVries, H., 22
Diatoms, 41, 55
Dinoflagellates, 40–41. *See* Protists
Dinomastigotes, 40, 41
Diploids, 101–102
Directed panspermia (pangenesis), 73–74
Direct filiation theory, 40–41
Dissipative structures, 78

Dissociation, 99–100
DNA (deoxyribonucleic acid), 21, 56, 67, 71
centriole-kinetosome, 47–48
of *Mycoplasma geniticulum*, 73
and origins of life problem, 73, 80–83, 84
ribosomal genes of mitochondria, 40
and symbiotic theory of origins of mitochondria, 42
Dobzhansky, Theodosius, 7–8, 23–24
Domains, 65, 68
Doolittle, Ford, 42
Down's syndrome, 102
Drosophila (fruit fly), 7–8, 23–24
Dyson, Freeman, 81

Ebola virus, 64
Ecosystems
colonization of, 106–111
defined, 105–106
recycling in, 94–95, 105–106, 119, 121
Ediacarans, 95
Egg cells
cytoplasmic factors outside nucleus, 22, 24–25

in fertilization process, 23, 52, 91–93, 98, 99–103
See also Sex
Eigen, Manfred, 81–82
Eldredge, Niles, 7, 8
Electron micrographs, 28
Electron microscope, 43, 56, 64
Embranchments, 59
Embryology, 23, 52, 73
Embryos, 52, 88, 91–92, 95, 100
Ephrussi, Boris, 19
Eubacteria, 43–44, 48, 65–66, 67
Eukarya, 65, 66
Eukaryotic cells, 7, 42, 45, 48, 57, 61
Eupatorium, 28
Eurodina, 97
Evolutionary theory
bacteria as basis of, 4, 53, 56, 64–65, 68, 70, 121
evolution defined, 24
natural selection and, 4, 8, 121, 124
punctuated equilibrium in, 8
role of symbiosis in, 6, 8
spirochete hypothesis for origins of cilia, 39–49
taxonomies based on, 59–68

Evolution by Association (Sapp), 19
Factors (Mendel), 19, 20–21
Family trees, 51–52
Famintsyn, A. S., 25
Fertilization process, 23, 52, 91–93, 98, 99–103
Fetuses, 91–92
Fisher, R. A., 23
Fitness, 18
Formaminifera, 60
Fossil record, 58, 59
 ecological crises and, 120
 microfossils, 65, 71, 72
 oldest animals in, 93
 oldest plants in, 107
 punctuated equilibrium in, 8
 virtual, 80
Fractal geometry, 126
Fruit fly (*Drosophila*), 7–8, 23–24
Fungal cells, 35–36
Fungal fusion, 107–108
Fungal symbionts, 5
Fungi, 56, 57, 68
 coevolution with algae, 107–109
 colonization of Earth, 107–111
 mold, 60–61, 62, 93
 mushrooms, 52, 61
 mycorrhizae, 5, 107–111
 sex among, 88, 91
 yeasts, 22, 28, 54, 61

Gaia hypothesis, 113–128
 acidity/alkalinity and, 120, 123
 basis of, 2, 118–119, 120, 123
 criticisms of, 125–126
 Daisyworld, 126–128
 development of, 115–119
 gases and, 115–117, 120, 121–123
 human role and, 115
 origin of term, 118
 popular culture and, 120
 proprioception and, 113–114, 126
 serial endosymbiosis theory (SET) and, 1–2
 "Strong Gaia," 124
 superorganisms in, 114–115, 123–126
 temperature regulation and, 117–118, 120, 123–125, 126–128
 "Weak Gaia," 123–124
Gene(s)
 in chloroplasts, 22
 location on chromosomes, 21, 22
 in mitochondria, 22
Gene mutations, 23
 through microbial symbiosis, 8

 through natural selection, 4, 8, 18
 X-rays and, 28
Gene sequencing, 6, 67, 73
Genetics, 17–18
 cell. *See* Cell genetics
 connection with chemistry, 24–25
 population, 18
Genus, 59
Geophysiology, 114, 118
Gilbert, Wally, 82
Glycine, 76
Golding, William, 118
Gonium, 97
Gould, Stephen Jay, 8
Gray, Michael, 42
Great Chain of Being, 3–4
Greeks, ancient, Great Chain of Being and, 3
Gupta, Radney, 43–44, 48

Haeckel, Ernst, 57, 59–60
Haldane, J. B. S., 23
Hall, John, 47
Hallucinogenia, 95
Halobacteria, 66
Haploid cells, 100–103
Harding, Stephen, 127
Hardy, G. S., 23
Harris, Betsy Blunt, 122

Hartman, Hyman, 43
Henneguy, L. F., 46
Henneguy-Lenhossek
    theory, 46
Heredity
    cytoplasmic
        inheritance, 27
    in nuclear
        organisms, 26–27
Heterotrophs, 83
Hinkle, Greg, 1–2
Hogg, John, 60–61,
    62
Homeostasis,
    117–118
Homo sapiens. See
    Human species
Human species
    classification of, 59
    dating of start of
        human life, 91
    fertilization and
        reproduction
        process, 91–93
    future as species,
        11–12
    and Gaia
        hypothesis, 115
    Great Chain of Being
        and, 3–4
    need for species
        recognition and,
        98–99
    sex among, 88
    tenure on earth, 91
Hutchins, Robert M.,
    16–17
Hydras, 10, 89
Hydrogen, 76, 83,
    116, 121
Hydrogen sulfide,
    121

Hypersea, 109–110,
    111
Hystrichosopheres,
    101

Independence,
    tendency to bind
        together and
        reemerge, 11–12
Individuality
    emergence of cell,
        10–11
    partners in origin of,
        34–36, 38
    symbiogenesis and,
        9
Inheritance of
    acquired
        characteristics,
        8–9, 27

"Jelly ball" bodies, 11
Joyce, Gerald, 82
Judeo-Christian
    tradition, Great
        Chain of Being
        and, 3

Kappa-killer
    inheritance
        pattern, 28
Keeble, J., 9–10
Kefir, 11–12
Kingdoms, 68
Kirshner, J., 124

Lamarck, Jean
    Baptiste, 8–9
Lamarckianism, 8–9,
    27

Leeuwenhoek,
    Antony van, 54,
        57
Lenhossek, Mihaly
    von, 46
Lichen, 108–109
Life Itself (Crick),
    73–74
Lignin, 110
Linnaeus (Carolus
    von Linné),
        58–59, 60
Liposomes, 79–80, 82
Long-chain
    molecules, 56
Lovelock, James E., 2,
    115–119,
    122–123, 124, 126
Luck, David, 47

Malloch, D. W.,
    107–108
Mars, 116
Mastigias medusoids,
    10
Material basis, 21
Mayr, Ernst, 66–67
McMenamin, Dianna,
    94, 109–110, 111
McMenamin, Mark,
    94, 95, 109–110,
    111
Meiosis, 26–27, 98
Meiotic sex, 88–89,
    90, 99–103
Memory, and origins
    of life problem,
    77, 79–80
Mendel, Gregor,
    19–21, 22, 23, 26

Merezhkovsky,
Konstantin S.,
25–26, 38, 43, 54
Metabolism, 24, 47
cumulative
metabolic chart of
living organisms,
79
in Gaia hypothesis,
123–124
inside cells, 82
and origins of life
problem, 73,
77–85
primary, 77–78
viruses and, 63–64
Methane, 83, 117,
121, 122
Methanogens, 117
Methyl chloride, 121
Methyl iodide, 121
Microbial mats,
69–70
Microbial symbiosis,
in fruit fly
research, 7–8,
23–24
Microbiology, 54
Microfossils, 65, 71,
72
Microscope
electron, 43, 56, 64
invention of, 54, 57
Microtubules, 42, 45,
47, 48–49, 102
Miller, Stanley L.,
76–77
Mills, Don, 81
Mirabilis jalapa, 28
Mitochondria, 6, 19,
25, 28–29

characteristics of,
39–40
direct filiation
theory of origins
of, 40–41
genes in, 22
non-symbiotic
origins of, 40–41
oxygen-breathing,
37–38, 45
symbiotic theory of
origins of, 42–49,
53
Mitosis, 26–27, 36,
44, 45, 46, 48, 98
Mitotic spindle, 42,
102
Mohammed pellets,
11
Molds, 60–61, 62, 93
Molecular biology, 6
Molecular symbiosis,
81
Molina, Mario, 116
Monera kingdom,
60–61
Monerans, 52–53
Morgan, Thomas
Hunt, 22, 23, 25
Morowitz, Harold J.,
71–72, 77–81,
83–85
Motility proteins, 47
Muller, Hermann J.,
23, 28
Multicellularity, 93
Mushrooms, 52, 61
Mutational load, 18
Mycology, 54
Mycoplasma
geniticulum, 73

Mycorrhizae, 5,
107–111

Naked genes, 28, 37
Natural selection, 4,
8, 121, 124
Neo-Darwinism, 18,
23, 124
Neo-Lamarckianism,
8–9
Neurons, 48
New individuals,
symbiogenesis
and, 9
New species,
formation of, 7–8
Nitrogen, 106, 116,
122
Nitrogen fixatives, 5,
106
Nucelocytoplasm,
34–36
Nuclear genetics
basic concepts of,
18–22
heredity in, 26–27
mitosis and, 26–27,
36, 44, 45, 46, 48,
98
origins of research
in, 19–22, 23
symbiogenetic
theory of origin of
cells with nuclei,
33, 42–49
Nucleic acids, 47
Nucleocytoplasm, 38

Oenothera, 28
Olendzenski,
Lorraine, 11, 122

On the Origin of Species (Darwin), 6

Ophrydium, 11

Orders, 59

Organelles, 19, 24, 25
bacteria as former, 34
chloroplasts, 19, 22, 29, 36–38, 40–41, 53
cilia. *See* Ciliate genetics
electron micrography and, 28
mitochondria. *See* Mitochondria
plastids. *See* Plastids
theory of origins of, 34–49

Origin of Sex, The (Margulis and D. Sagan), 89

Origins of life problem, 70–85
approaches to, 70–71
Belousov-Zhabotinsky system and, 78–79
directed panspermia (pangenesis) and, 73–74
direct experiments on, 76–77, 79–80
DNA and, 73, 80–83, 84
memory and, 77, 79–80
molecular symbiosis and, 81

Harold Morowitz and, 71–72, 77–81, 83–85
primary metabolism and, 77–78
primordial soup and, 76, 80
properties of minimal bacterial life and, 70–71
spontaneous generation and, 74–76
Swaziland microspheres and, 72
thermodynamics and, 77–79, 82–85

Oxygen, 35–38, 44, 45, 83, 94, 116, 117, 121

Paleobiology, 71

Paleozoic Era, 106

Pandorina, 97

Pangenesis (directed panspermia), 73–74

Paramecium, 27, 28

Parents, symbiogenesis and, 9

Pasteur, Louis, 57, 74–75

Perpetual motion machines, 78–79

Phaeophytes, 41

Phosphalites, 95

Phosphorus, 109

Photoautotrophs, 83–84

Photosynthetic bacteria, 9–10, 22, 36–37, 55, 109

Phyla, 59

Pikaia, 95

Pirozynski, K. A., 107–108

Plachobranchus, 10

Plankton, 60

Plant-animals, 9–10

Plant-Animals (Keeble), 9–10

Plants, 52–53, 54, 57, 68
colonization of Earth, 107–111
evolution of, 95–98, 108–111
oldest fossils of, 93
sex among, 88–89, 99
trees, 110–111

Plastids, 6, 25, 28–29, 38
characteristics of, 39–40
direct filiation theory of origins of, 40–41
ribosomal genes of, 40

Playmonas cells, 9–10

Pliny, 58

Pontecorvo, Gino, 25

Population genetics, 4, 8, 18, 23

Population overgrowth, 128

Precambrian, 93

Prigogine, Ilya, 78

Primary metabolism, 77–78

Primates, Great Chain of Being and, 3–4

Primordial soup, 76, 80

Prokaryotic cells (monerans), 7, 52–53, 57, 61, 66, 67

Proprioception, 113–114, 126

Protein creatures, 81

Proteins, 42, 47, 56
  and origins of life problem, 73, 80–81
  See also Amino acids

Proteobacteria, 38

Protists, 28
  abortive cannibalism among, 89, 99–103
  defined, 62
  in other taxonomies, 54–55, 57, 61, 62
  sexual fusion of, 91, 92
  theory of origins of, 34–36, 40–41, 46, 61, 65

Protoctists, 36, 44, 52–53, 57, 68
  abortive cannibalism among, 89, 99–103
  cloning and, 93
  evolution of death among, 90

as haploids, 101
  mitochondria in, 38
  nature of, 65, 67
  in other taxonomies, 54–55, 60–61
  sex among, 84, 88, 98–103

Protomating, 102–103

Protozoa, 54–55, 60–61, 62

Pteridinium, 95

Punctuated equilibrium, 8

Radiolaria, 60

Ramanis, Zenta, 47

Ray, John, 58

Recycling, 94–95, 105–106, 119, 121

Rhodoplasts, 41

Rhynia, 109–110

Rhynie chert, 107, 108

Ribonucleotides, 82

Ribosomes, 40, 67, 80–81

Ribozymes, 82

Ris, Hans, 28

RNA (ribonucleic acid), 56, 65–66, 67
  and origins of life problem, 73, 80–83, 84

Rosenbaum, Joel, 48

Rowland, Sherwood, 116

Rumen, 122

Ryan, Francis, 25

Sagan, Carl, 16–17, 22, 24

Sagan, Dorion, 2, 52–53, 89, 125

Sager, Ruth, 25

Sapp, Jan, 19, 21

Schistosoma, 103

Schneider, S., 124

Schwartz, Karlene V., 56–57, 61–68

Serial Endosymbiosis (Margulis), 29

Serial endosymbiosis theory (SET)
  alternative hypothesis to, 40–41
  basis of, 6
  central idea of, 37
  concept of "serial" in, 34
  current version of, 30–32, 34–49
  earliest complete statement of, 29–30
  early experimental contributions to, 30
  "extreme SET," 39–49
  Gaia hypothesis and, 1–2
  origins of organelles in, 34–49
  spirochete hypothesis for origins of cilia, 38–49
  visual representations of, 31, 52–53, 129

Sex
  as abortive
    cannibalism, 89,
    99–103
  among animals,
    88–89, 99
  among bacteria,
    87–89
  fertilization process
    and, 23, 52,
    91–93, 98, 99–103
  among fungi, 88, 91
  meiotic, 88–89, 90,
    99–103
  among plants,
    88–89, 99
  programmed death
    and, 90
  among protists, 91,
    92
  among protoctists,
    84, 88, 98–103
  symbiosis versus,
    89–90, 103
Silent Spring
  (Carson), 116
Simpson, George
  Gaylord, 17
Slime molds, 60–61,
    62, 93
Smith, David C., 46
Sonneborn, Tracy,
    27, 28
Species, 11, 59
  mutability of, 20
  symbiogensis of, 6
  symbiosis as basis
    of, 6
Species recognition,
    98–99
Spemann, Hans, 23

Sperm-egg unions,
    23, 52, 91–93, 98,
    99–103
Sperm tails, 42
  activity of severed,
    47
  microtubules of, 48
Spiegelman, Sol, 81
Spirochete
  hypothesis
  and evolution of
    meiosis, 102–103
  for origins of cilia,
    38–49
Sponges, 60
Spontaneous
  generation, 74–76
Star Trek, 105, 111
Stentor, 23
Stentor coeruleus, 90
Sturtevant, A. H., 24
Sulfate, 109
Supermolecules, 81
Superorganisms,
    114–115,
    123–126. See also
  Gaia hypothesis
Swaziland
  microspheres, 72
Symbiogenesis, 57,
    68
  acquisition of entire
    organisms
    through, 8–9
  bodies formed by,
    89
  cytoplasmic genetics
    as central area in,
    20
  defined, 6, 33
  in distinguishing
    nucleated- from

  bacterial cells,
    42–43
  emergence of new
    life forms
    through, 9,
    107–111
  in habitation of
    Earth, 107–111
  as neo-
    Lamarckianism,
    8–9
  origins of concept
    of, 38
  origins of organelles
    and, 34–49
Symbionticism, 6,
    53–54
Symbiosis
  in colonization of
    Earth, 106–111
  defined, 1–2, 5, 33
  microbial. See
    Microbial
    symbiosis
  misuse of term, 81
  molecular, 81
  origins of term, 33
  plant-animals and,
    9–10
  prevalence of, 5
  punctuated
    equilibrium and,
    8
  role in evolution, 6
  sex versus, 89–90,
    103
  as source of
    evolutionary
    novelty, 6, 8, 9,
    20, 33
Szostak, Jack W., 82

Tartar, Vance, 23
Taxonomies, 51–68

author's revision of previous, 52–53, 56–57, 61–68
confusing terms in, 55
defined, 51
evolution-based, 59–68
functions of, 51–52
pre-evolutionary, 57–59
problem terms and, 54–55

Taylor, Max, 29, 30, 39, 40, 48
Temperature regulation, Gaia hypothesis and, 117–118, 120, 123–125, 126–128
Thermoacidophil bacteria. See Archaebacteria
Thermodynamics, 77–79, 82–85, 119
Thermoplasma, 43

Tissue differentiation, 98
Trees, 110–111
Tribrachidium, 95
Truffles, 110
Tubulin, 42, 48–49

Undulipodia, 47, 48
University of California at Berkeley, 26–29
University of Chicago, 15–18, 22–24
University of Wisconsin at Madison, 17, 24–26, 28
Urey, Harold C., 76–77

Vernadsky, Vladimir, 110
Virtual fossils, 80
Viruses, nature of, 63–64, 71
Volvox, 97–98

Wallin, Ivan E., 6, 25, 53–54, 55
Waste, 94–95, 105–106, 119, 121, 128
Watson, Andy, 126
Weismann, August, 23
Whittaker, Robert H., 61, 62, 65
Wilson, E. B., 25
Wilson, E. O., 114
Woese, Carl R., 48, 65–66, 67

X-rays, mutations from, 28

Yeast, 22, 28, 54, 61

Zea (corn), 28
Zygospores, 101
Zygotes, 73, 88, 91

ABOUT THE AUTHOR

Lynn Margulis, Distinguished Professor in the Department of Geosciences at the University of Massachusetts at Amherst, has been a member of the National Academy of Sciences since 1983. She is best known for her pathbreaking work on the bacterial origins of cell organelles and for her collaboration with James Lovelock on Gaia theory. Her previous books include *Symbiosis in Cell Evolution; Five Kingdoms* (with K. V. Schwartz); and (with Dorion Sagan) *Origins of Sex; Garden of Microbial Delights; What Is Life?; What Is Sex?*; and *Slanted Truths: Essays on Gaia*.